Inkscape by Example

A project-based guide to exploring the endless features of
Inkscape and upgrading your skills

István Szép

BIRMINGHAM—MUMBAI

Inkscape by Example

Copyright © 2023 Packt Publishing

Group Product Manager: Rohit Rajkumar

Publishing Product Manager: Ashitosh Gupta

Senior Editor: Keagan Carneiro

Content Development Editor: Abhishek Jadhav

Technical Editor: Simran Ali

Copy Editor: Safis Editing

Project Coordinator: Sonam Pandey

Proofreader: Safis Editing

Indexer: Tejal Daruwale Soni

Production Designer: Joshua Misquitta

Marketing Coordinator: Nivedita Pandey

First published: December 2022
Production reference: 1141222

Published by Packt Publishing Ltd.
Livery Place
35 Livery Street
Birmingham
B3 2PB, UK.

ISBN 978-1-80324-314-6

www.packt.com

To my wife, Erika, who is an absolute champ, and who tolerated and supported me while writing this book! And to my son, Ádám, who will be born in 2 weeks, thus forcing me to finally finish writing this book!

István Szép

Contributors

About the author

István Szép is a graphic designer, illustrator, and design teacher. His goal is to inspire others by sharing his knowledge about visual design and life and work as a freelance designer. His tool of choice is Inkscape, an open source graphic design program, which is what he uses most for his design work.

István earned his master's degree as a visual communication and multimedia teacher in 2009 from **MOME (Moholy-Nagy University of Art and Design)**. Since then, he's worked as an illustrator and graphic designer for hundreds of clients, while teaching both live and online classes to thousands. He currently lives and works in Budapest with his wife, Erika, who is also a graphic designer.

About the reviewer

Chris Hildenbrand has been a pixel-pusher, vector-bender, and quad-turner since the days of the Commodore 64. He went on to illustrate games for the Commodore Amiga, Atari ST, and early PC. A long line of Flash games followed. Among them was *Heli Attack 3*, created by Chris Rhodes. A vast number of mobile and handheld game projects contain his pixel- and vector-based art.

Some Xbox Live Arcade games, a long list of iPhone/iPad games, and many game development templates later, he started writing and recording tutorials geared at programmers with little to no art skills. Inkscape was the ideal tool for the task – free and open source. *Game art for programmers* was born and turned into *2Dgameartguru* nearly 10 years later.

Table of Contents

7

Combine Inkscape and Other (Free) Programs in the Design Workflow

8

Pro Tips and Tricks for Inkscapers

9

Conclusion   195

Index   201

Other Books You May Enjoy   208

Preface

Welcome to *Inkscape by Example*! Inkscape is a popular vector graphics program. It is free and has a wide range of tools and a great online community. Also, it is easy to use on a basic level, but there is a steep learning curve if you want to use it professionally. This book is here to help with that. Just follow my example!

Who this book is for

This book showcases several projects suitable for graphic designers, UI designers, illustrators, art directors, digital artists, and other creatives looking to elevate their Inkscape skills. If you are an application developer who knows the basics, but want to get out more of the application, and be able to create vector graphics for various professional or DIY projects, this book offers the next step. The aim of the projects in this book is to help you build all the skills and routines you need to master Inkscape and create high-quality vector graphics on your own.

What this book covers

Chapter 1, Inkscape Is Ready for Work – Design a Business Card as a Warmup!, answers the question is Inkscape the relevant tool for you? In the first chapter we will list and define the strengths of Inkscape. You will then learn about the new tools and features in Inkscape 1.0. and 1.2. Finally, you can test your current Inkscape skill level with a fast warm-up exercise!

Chapter 2, Design a Clever Tech Logo with Inkscape, looks at designing a logo, which is a common task faced by a professional designer. Inkscape gives us all the flexibility we need to create a smart logo! In this chapter, we will create a simple logo for a tech company, using the different shape tools and Boolean operations in Inkscape. Then, we will create logo mutations via *duplication* and export the logos into different formats.

Chapter 3, Modular Icon Set Design with the Power of Vector, discusses how while creating a single icon is usually a basic task, creating a whole set of icons can be a real challenge! We will use the power of vectors and create a modular set of nine icons for any theme! The flexibility of Inkscape will help you set and maintain your own design ruleset throughout the process.

Chapter 4, Create Detailed Illustrations with Inkscape, discusses how while basic vector illustrations are a good way to learn Inkscape, to provide design services for a client or reach your artistic goals as an illustrator, you need to know the tools and rules to create more detailed drawings in Inkscape. The project in this chapter focuses on the workflow of creating a business illustration with many details and elements.

Chapter 5, Edit a Photo and Create a Hero Image in Inkscape, explores how while Inkscape is a vector graphics software, it has tools for photo editing too! You can use these tools and filters to your advantage while creating layouts or trace photos to use them as scalable vector elements in your designs. In this project, we will clip and trace photos to use them as story elements in a website header.

Chapter 6, Flexible Website Layout Design for Desktop and Mobile with Inkscape, as any current website has to behave responsively, discusses how there is no better tool to create flexible layout mockups than a vector. In this chapter's project, you will create a simple website layout for three views, using the elements we created in the previous chapters. We will focus on using an effective workflow and learn how to rescale and reshape elements for different screen sizes. We will also learn about the SVG format and how it is great for illustrations targeted for the web!

Chapter 7, Combine Inkscape and Other (Free) Programs in the Design Workflow, discusses how while Inkscape is a great vector graphics application, you should use it in combination with other programs for a really efficient workflow. In this chapter, you will get to know a few examples of free software that work great with Inkscape. In the first example, you will enhance your Inkscape illustrations with Krit. In the second, you will learn about Scribus, which is used to create desktop publishing with crisp vector elements imported from Inkscape files. The third part is about Inkscape and DragonBones, to give you a taste of 2D illustration possibilities. The last example in this chapter is Blender with SVG files, and how to turn them into 3D fast!

Chapter 8, Pro Tips and Tricks for Inkscapers, is about situations you might encounter using Inkscape, and the solutions and workarounds for them. CMYK? SVG? EPS? PDF? XML? We share tips about navigating your SVG file and exporting it into different formats, and how to use LPEs to your advantage and work faster than ever before!

Chapter 9, Conclusion, provides a short summary of topics we covered in this book, the type of projects we completed. This chapter concludes everything we covered in this book.

To get the most out of this book

You can follow along with the chapters individually, but the projects are also loosely built on each other. They are also ordered in level of difficulty, from shorter projects to more complex ones. Each chapter shows an example of what you can create with Inkscape, and all the possibilities you have using the program!

By the end of this book, you will be competent in creating your own solutions to any projects with Inkscape and tackle all the tasks a vector designer should be able to do.

Software covered in the book	OS requirements
Inkscape 1.1 and 1.2	Windows, macOS, or Linux (any)
Scribus 1.5.7	Windows, macOS, or Linux (any)

Download the supporting files

You can download the supporting files for this book from GitHub at `https://github.com/PacktPublishing/Inkscape-by-Example`. If there're any updates to the files, they will be updated in the GitHub repository.

We also have other code bundles from our rich catalog of books and videos available at `https://github.com/PacktPublishing/`. Check them out!

Download the color images

We also provide a PDF file that has color images of the screenshots and diagrams used in this book. You can download it here: `https://packt.link/muZqP`.

Conventions used

There are a number of text conventions used throughout this book.

`Code in text`: Indicates code words in text, database table names, folder names, filenames, file extensions, pathnames, dummy URLs, user input, and Twitter handles. Here is an example: "Select and copy the user shape into a new `.svg` document."

Bold: Indicates a new term, an important word, or words that you see onscreen. For instance, words in menus or dialog boxes appear in **bold**. Here is an example: "Feel free to create your own colors, using the color wheel in the **Fill and Stroke** window."

> **Tips or important notes**
> Appear like this.

Get in touch

Feedback from our readers is always welcome.

General feedback: If you have questions about any aspect of this book, mention the book title in the subject of your message and email us at `customercare@packtpub.com`.

Errata: Although we have taken every care to ensure the accuracy of our content, mistakes do happen. If you have found a mistake in this book, we would be grateful if you would report this to us. Please visit www.packtpub.com/support/errata, selecting your book, clicking on the Errata Submission Form link, and entering the details.

Piracy: If you come across any illegal copies of our works in any form on the Internet, we would be grateful if you would provide us with the location address or website name. Please contact us at copyright@packt.com with a link to the material.

If you are interested in becoming an author: If there is a topic that you have expertise in and you are interested in either writing or contributing to a book, please visit authors.packtpub.com

Share your thoughts

Once you've read *Inkscape by Example*, we'd love to hear your thoughts! Scan the QR code below to go straight to the Amazon review page for this book and share your feedback.

https://www.amazon.com/dp/1803243147

Your review is important to us and the tech community and will help us make sure we're delivering excellent quality content.

Download a free PDF copy of this book

Thanks for purchasing this book!

Do you like to read on the go but are unable to carry your print books everywhere?

Is your eBook purchase not compatible with the device of your choice?

Don't worry, now with every Packt book you get a DRM-free PDF version of that book at no cost.

Read anywhere, any place, on any device. Search, copy, and paste code from your favorite technical books directly into your application.

The perks don't stop there, you can get exclusive access to discounts, newsletters, and great free content in your inbox daily!

Follow these simple steps to get the benefits:

1. Scan the QR code or visit the link below:

https://packt.link/free-ebook/9781803243146

2. Submit your proof of purchase

That's it! We'll send your free PDF and other benefits to your email directly.

1

Inkscape Is Ready for Work – Design a Business Card as a Warmup!

Inkscape is free and open-source software that's been in development for a long time. So, in this first introductory chapter, you will get an overview of the current state of the application. Then, you will design a business card as a self-assessment exercise to learn your true Inkscape level and refresh your skills. By the end of this chapter, you will not only learn the aim and method of this book but also know which Inkscape skills you need to improve on!

In this chapter, we're going to cover the following main topics:

- Welcome to Inkscape 1.2 – and the next level of Inkscape!
- A short self-assessment exercise
- Designing a clean business card step by step

Welcome to Inkscape 1.2 – and the next level of Inkscape!

As you know, Inkscape is free. You do not have to pay to use it and you can create with it without any restrictions.

Being free means it is an accessible vector graphics solution for students, amateurs, and hobbyists, but it also means that the developer team is working on it as volunteers. This makes the development much slower than that of the profit-oriented development process of Adobe Illustrator, for example.

The development of the predecessors of Inkscape started around 1998, and the first Inkscape version appeared in 2003. The development aimed to create a well-rounded open-source SVG editor. During those years, the Inkscape developer team kept adding feature after feature so that this software could grow into its latest version.

Inkscape 1.0 was officially released in May 2020. This was a huge milestone, a stable release with many new tools and design solutions. Since then, we had Inkscape 1.1 and 1.2, and the version numbers will steadily grow with every new feature added.

For a long time, Inkscape was considered a tool for hobbyists, but this mindset started to shift. More and more, professional designers use Inkscape as their main tool. After a humble start, the developers continuously grew the toolset and the list of features of the program so that Inkscape is ready to compete with well-established vector editor programs on the market (such as Adobe Illustrator).

Of course, there is still room for improvement – user requests for new tools and features are popping up every day. But even with some shortcomings, Inkscape is ready for high-level use, and it is a relevant design tool today.

The following sections will address most of the strengths Inkscape developed during recent years.

Stability and supported operating systems

You can download a stable native version of the program for Windows or Linux, and the certified macOS version is also available. Inkscape running on Windows and Linux also received a performance boost with version 1.0.

Everything will run smoothly while you are working with hundreds of objects, applying filters, or using live path effects that are usually demanding on hardware. Inkscape also has a good bug-tracking system in place, and the Inkscape user community is very involved in reporting them to be fixed.

Professional tools and features

Inkscape has come a long way, and we can safely state that most of the things you can create with Illustrator can be created with Inkscape too! There are tools that designers used in other programs and wanted to have in Inkscape too or tools that would make working with Inkscape simply easier. Here are a few examples from the latest builds.

There is a whole new dialog window for **Live Path Effects** (**LPE**) to make dynamic changes to any path. This is the best tool if you are illustrating in Inkscape and want to render **Perspective** or align a **Pattern Along a Path** to spare time repeating shapes rendered along a path. Alternatively, you can apply **Power Stroke** to draw line art with an organic-looking ink line effect.

In the latest stable versions, these effects became fast and reliable, ready to make your life easier! You will learn more about them in *Chapter 8, Pro Tips and Tricks for Inkscapers!*:

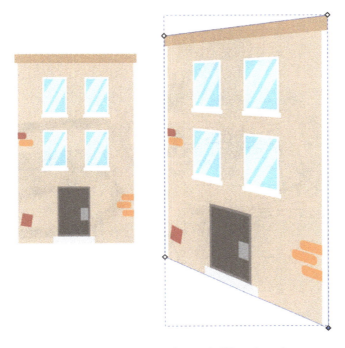

Figure 1.1 – Perspective live Path Effects in action

The **Gradient mesh** tool got introduced in version 0.92 and is now a stable solution for creating illustrations with gradients built up using multiple colors with a grid and grid points. It takes some time to get used to, but it is a professional way to create subtle coloring and gradients in your design work.

The **Measurement** tool is a ruler that you can use to measure length in Inkscape. Whether it's the distance between objects or nodes along a path, the ruler gives you a proper measurement.

Another new feature is **Split Mode**, which you can find under the **View** menu. Using it, you can watch your artwork simultaneously in two display modes – normal mode on the left, and **Outline** on the right. This can help you identify and edit overlapping objects easier on the fly. **Outline overlay** mode is another great view mode, where an outline of the objects is displayed while showing their real colors as well:

Figure 1.2 – Editing objects in a complex illustration using Split Mode

Inkscape 1.2 marked the introduction of the **Page** tool, which manages page setup and allows the usage of multiple paged Inkscape documents! This is an important step toward professional Inkscape usage, since working with multiple pages is important for desktop publishing, user interface design, or general graphic design.

Also, with Inkscape 1.1 came new export formats such as WEBP, and version 1.2 introduced revamped **Batch export** capabilities that make working with Inkscape even more convenient. These features allow you to export groups or selected objects in different versions in a few clicks.

One long-awaited tool is **Shape builder**, which will arrive in Inkscape 1.3, but you can try it out already in the developer versions. It allows you to merge and cut shapes and create new shapes more conveniently.

Quality resources and tutorials

This one is not a direct feature of Inkscape, but rather an indirect result of the hard work of the Inkscape developer team and the user community mentioned previously. After years of listening to the community and adding new features and solving bugs, the Inkscape team convinced professionals and educators to use Inkscape as their main vector design tool.

This resulted in countless high-quality portfolio examples and learning materials to help newcomers find and adopt Inkscape. This growth in available pro users will also demand even more from the developer team, thus making future development faster.

CMYK handling and creating print-ready files

If you want to work with a printer, most of the time, you will be asked for a vector file exported in the CMYK color space. As a designer, this will happen eventually, so it is a very common feature all graphic design programs must have. And I will be honest, Inkscape falls short here.

The program has CMYK support but can't display the set CMYK color in RGB as Adobe products can. This means that if you want to export CMYK files from Inkscape, and want to see the proper matching colors on your monitor, you will need to use a third-party solution at the moment.

However, this is possible via extensions or using a program that can output proper CMYK, such as Scribus. I will explain these methods in *Chapter 7*, *Combine Inkscape and Other (Free) Programs in the Design Workflow*, so that you can try them for yourself!

In my opinion, these strengths perfectly answer the question: is Inkscape good enough for professional everyday use? Yes, it is! It is stable and resourceful and as you will see while completing the projects in this book, it is a very versatile design program.

I hope that after reading the previous sections, you can also see that Inkscape is becoming a professional tool and that it is worth the effort to learn to use it at a higher level.

A short self-assessment exercise

Are you ready for a warmup to refresh your Inkscape knowledge? To reach a higher level in any skill, knowing your current level is important. The goal of the following test exercise is to give you a realistic view of your Inkscape skills.

This short test is self-graded, so finish the exercise on your own first, then answer the questions of the evaluation. After that, you are free to look into my method to solve this task on the following pages, but for the best learning experience, please finish the task first on your own!

Your task: copy this business card design in the best possible way you can!

Figure 1.3 – Recreate this business card design in Inkscape

Here are a few hints and criteria:

1. Create a card that is standard size 90*50 mm (or 3.5"x2").

2. Add the two background shapes on the left and the text on the right.

3. Work with text and color and use gradients.

4. Save the card as a PDF file.

> **Note**
>
> We don't need to add bleed to our business card, and it doesn't have to be print-ready either, as this is just a design exercise.

Are you ready? Start!

Evaluating your current Inkscape skill level

Now that you've finished this assignment, it is time to evaluate it. Give yourself 1 point for each task you completed with ease:

1. You drew a rectangle and set its size to 90mm by 50 mm.

2. You set the background color of the card and used the **Gradient** tool.

3. You added the text and used different font sizes.

4. You created the two shapes on the left using the **Path editor** tool and/or using path Boolean operations.

5. You saved the card as a PDF file of the correct size.

Based on your total points, this is your current skill level:

- **4-5 points**: You have used Inkscape before and have practical design knowledge! You can move forward to the next few chapters and learn even more about the professional workflow with Inkscape!

- **2-3 points**: You know the basics and can find your way around the surface of Inkscape but need more practice. The projects in this book will help you develop those skills further while setting you up for a challenge.

- **0-1 point**: To get the maximum out of this book, you will need to learn the basics first and get familiar with Inkscape. I've added some links here that will help you with the basics of Inkscape. Once you feel comfortable with your skill level, read on to become an Inkscape pro!

If you want to take a look at how I designed the card, read through the following section for my method.

Designing a clean business card step by step

Even if you scored high on the previous design test, feel free to read the following tutorial.

We will follow the same structure in all of the projects in this book, going step by step, focusing on actionable tasks, and learning by example. I will explain all the tools that are necessary for the task at hand so that you can keep the focus on learning what you need. We will cover almost all the possible tool combinations in Inkscape, but object-oriented, not tool by tool. So, here is how I created the business card design.

Step 1 – Creating a rectangle and setting it to the correct size

Use the **Rectangle** tool and draw a simple rectangle. The color does not matter for now; we will change that later. Set the size of the rectangle by selecting it and adding the width (**W**) and height (**H**) in millimeters in the Tool controls bar. If the unit is not set to millimeters, click on the units and select the proper unit for this project, as shown in *Figure 1.4*:

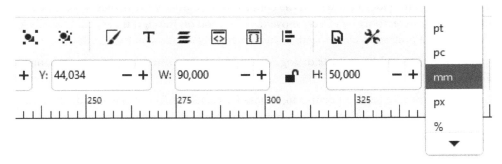

Figure 1.4 – Setting the width and height on the Tool controls bar and applying millimeters

Alternatively, you can navigate to **File | Document properties** and change the **Display** units to **mms** there! This way, all the tools and rulers will be in the correct unit.

Step 2 – adding text to the card

We will write the text first. As you need to set a visual hierarchy of the text on the card, you will need to apply different font sizes and font weights to the blocks of text. Text is just an object like any other object in Inkscape. You can move it, scale it, transform it, and recolor it easily.

Let's learn how to do this:

1. Click on the card using the **Text** tool and type the name.
2. Select the text and choose a font and set a **Font size** of **18** pixels to it. Also, set **Font weight** to **Semi-bold** or **Bold** to make the name stand out:

Figure 1.5 – Setting the font's weight and size

3. Now, write the title under the name using the **Text** tool. Inkscape will remember the last settings of the **Text** tool, so the font, weight, size, and color will remain the same.

4. Resize the title text so that it's smaller than the name on the card. You can also reduce the weight of the chosen font if necessary.

5. Then, as the third block of text, add the contact information of the owner. We created three blocks of text here, so we can freely move them around on the card if needed. I usually keep the name and profession close together, and the contact details a bit further down.

6. Open the **Align and Distribute** dialog window from the **Object** menu at the top, or press *Shift + Ctrl + A*.

7. In the dropdown list labeled **Relative to**, choose **Last selected**. This will make sure that the text objects will be aligned with each other, not to the edge of the card or the page.

8. Now, select the three text objects and align their *left edges*. This will visibly align all the text to the left, but in reality, the text objects will be aligned with each other, not the text itself in the objects:

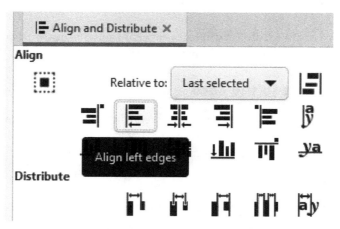

Figure 1.6 – Aligning the left edges of the text objects

9.   After the text is done, draw a thin rectangle as a horizontal separator between the two parts of the text.

> **Tip**
>
> It is smart to make the name and the role on the business card more visible than the other lines of contact information on the card. As the word suggests, a *business card* has an important role: to show the receiver what the profession of the owner of the card is. In my native language, we call such a card a *name note* (névjegy), which highlights the other function of the card: to clearly state the name of the owner.

Step 3 – creating the shapes on the left of the card

There are two methods you can use to create these shapes. Both are equally good for what we want to design here, and you are free to choose any of the two methods to continue. I usually use the one that I feel is fast and convenient for the project at hand.

We will practice both of them in the upcoming projects again and again, as they are the base of building custom shapes in vector.

Method 1 – using Boolean operations to create the shapes

Boolean operators are basic logical operations that help you merge and cut shapes and objects to create new shapes as you need. It is an effective method and very logical.

Here is how you can use this method in this case:

1.   Select the card and duplicate it by pressing *Ctrl + D*. Let's call the card Rectangle1. We used duplication here because that involves creating a copy of an object exactly above the original object so that the edge of the new shape will be in position with the original card. This new rectangle is named Rectangle2.

2.   Next, you must use the **Rectangle** tool and draw a slightly bigger rectangle (named Rectangle3) above the duplicated rectangle (Rectangle2). If you need to, set any color to these rectangles so that you can differentiate them.

3.   Rotate Rectangle3 at an angle, and take care that it is completely covering the right half of Rectangle2:

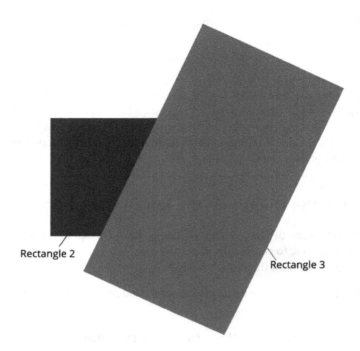

Figure 1.7 – These are your two rectangles overlapping

4.   Now, select both `Rectangle2` and `Rectangle3` and click on **Path | Difference** in the menu.

This command is used to cut off the shape on the top from the shape under it. It only works if two objects are selected. In this case, `Rectangle3` is cut to the size that is half of `Rectangle2`, creating the shape we wanted!

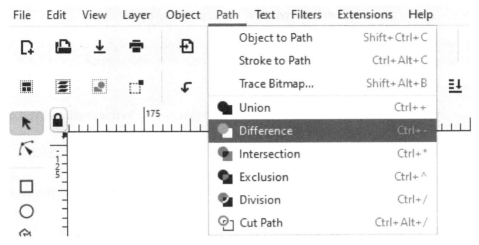

Figure 1.8 – Using the Difference path operation from the top menu

5. Now, select this shape and duplicate it by pressing *Ctrl + D*.

6. Select this new duplication and flip it vertically by pushing *V*! Now, you have both of the shapes on the left, and they are in perfect position! See *Figure 1.9* for reference:

Figure 1.9 – The state of the card design with the two shapes in place

---

Tip about learning the Boolean operations

There is a small icon before every Boolean operation in the **Path** dropdown list. If you are not sure about the function of the operation, just take a peek, it's ok!

Method 2 – using the Path editor tool to create the shapes

The **Path editor** tool is a very important tool that can take a beginner's design to another level. With it, you can shape elements of your design by adding, removing, or modifying nodes to the path.

This is how you can use it to design your business card:

1. Select the card and duplicate it with *Ctrl + D*. Set a different color to this duplicated rectangle.

2. Now, select this new rectangle and click on **Path | Object to path** or use the *Shift + Ctrl + C* hotkeys! This will turn the rectangle object into a path that you can edit with the **Path editor** tool.

3. Select the tool, click on the rectangle, and move the bottom-right node to the left while holding the *Ctrl* key. This will move the node on a horizontal line, thus keeping the edges of your shape aligned with the original card:

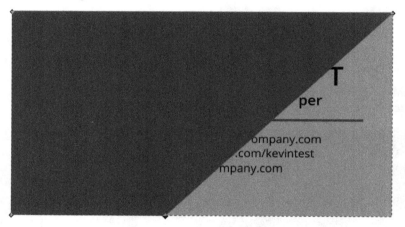

Figure 1.10 – Using the Node tool and moving this node to the left

4. Move it about two-thirds of the length of the card.

5. Now, select the node on the top right and move that to left as well, even further so that it creates a diagonal line with an approximate angle of 60 degrees.

6. Now, select this shape and flip it vertically by pushing *V*! Now, you have both of the shapes on the left, and they are in perfect position! (See *Figure 1.9* for reference).

Step 4 – setting colors and gradients

I usually don't focus on colors until the shapes in my design are almost set. I just use any placeholder colors at first and work on the shapes, proportions, and positions in my design. We have several objects to color in this case: the card, the text, the horizontal bar, and the two diagonal shapes on the left.

First, the background. Where there is text, you have to think about legibility. So, pick a background color that you will set to the card base, then a text color that is easy to read on that background. I used a dark blue for the text as it stands out against the light gray of the card:

1. Select the card and use the **Gradient** tool to apply a gradient that has a darker color on the left, creating a loose drop-shadow effect for the shapes on the left. To keep the design together, that darker side should be based on the same dark blue that is the color of the text.

2. Select the gradient you just created and, with the **Eyedropper** tool, pick the color of the text.

3. Then, grab the handler of the gradient and move it further, even out of the card if needed, to create a smooth gradient with a darkening effect:

Figure 1.11 – Applying a fine gradient to the card

4. Now, select the bottom of the two diagonal shapes and, while selected, use the **Color picker** tool. Again, pick the blue color of the text.

5. Apply a gradient to this shape as well from blue to a darker blue. To make the card a bit more interesting, we will color the other shape red, then add a gradient from red to blue. If you extend the gradient, it will go from red to dark purple, creating a nice effect and making the shapes pop out a bit.

6. Finally, select the horizontal bar between the title and the contact information, and set it to the same red. There's no need to add a gradient here – just some red to break the blues. Once you have finished all the steps, you should have a card that looks like the one shown in *Figure 1.3*!

> **Tip about gradients**
>
> Gradients are nice. But as in many cases, less is more. Try to use gradients only when the image benefits from them. I tend to use subtle gradients to create drop shadows, emulate lighting, or create color contrast. Ask yourself: do I need this gradient here?

Step 5 – grouping the elements and exporting the card

At this point, your card looks like the one I created, but the final task was to export it into a PDF file. Inkscape doesn't have multiple artboards, but one huge artboard to show. When you start up the program, the default is an A4 page in the middle.

As our card is much smaller, you could start by adding the size of the document. But what I usually do is the opposite. I create the design in a given size and, as a final step, resize the document around it:

1. Select all the elements of your business card and **Group** them from the right-click menu or by pressing *Ctrl + G*.

2. Now. with your group still selected, open the **File | Document properties** dialog and, under the page size, choose **Resize content to Page or selection**. Alternatively, use *Shift + Ctrl + R* for the same resize effect without even opening the document properties:

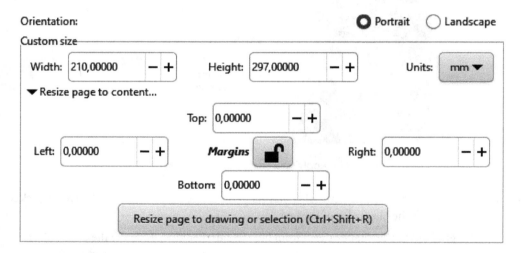

Figure 1.12 – Resizing the page to the group you've selected

Now your page is resized to the exact size of your business card!

3. To save it as a PDF, just save it from the file menu. After choosing the .pdf extension, Inkscape will show you the **Portable Document Format** pop-up window.

4. Here, you have to select **Output page size: Use document's page size**. This will save the PDF file as the size of the business card you designed:

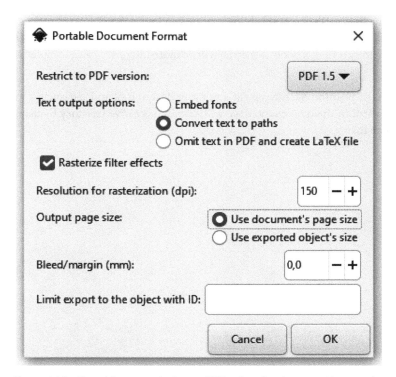

Figure 1.13 – Exporting your design to PDF, using the document's page size

This is how we designed the business card in Inkscape. First, drew the base of the card, then added the text and some decorative elements, and finally exported it as a PDF file. How was this workflow compared to your method? Did you learn something new? How would you do things differently?

This book shares the workflows I use as a designer working with Inkscape, but as you saw, there is not one ultimate solution. There are always different methods that utilize different tools. The more tools and methods you know, the bigger your design toolbox, and the easier it is to find an efficient solution for a visual problem.

Summary

This concludes this first short chapter in this book. After the introduction, we took a closer look at the current state of Inkscape and saw that after years of constant development, this awesome program offers fast and stable versions on all major platforms, has a growing community with plenty of resources, and is ready for daily design work – even for professionals.

Then, you met your first challenge in this book. I hope you found the self-assessment exercise useful and that you now know your real Inkscape skill level. And I also hope it was setting your expectations for the upcoming projects in this book as they all will be more interesting and complex than this short warmup exercise!

After establishing the position of Inkscape as a relevant design tool and measuring your skills, it is time to move forward! In the next chapter, we will dive into creative territory by using Inkscape to design a clever tech logo.

2

Design a Clever Tech Logo with Inkscape

Logo design is one of the most common tasks for graphic designers, so this will be the first in our chain of projects. In this chapter, we will create a logo for a tech company. We will start with designing the logo. Then, in the following chapters, we will look at icon design and illustrations and banner image design, and then combine all these into a website design. During this project, you will learn about how to use paths, color, and text effectively in Inkscape to create a fresh logo!

In this chapter, we are going to cover these main topics:

- Using Inkscape for logo design
- Starting the logo design process with the usual first steps
- Implementing shape and color in logo design in Inkscape
- Adding text to your logo and modifying letters for more personality
- Creating logo mutations in various formats for different usages

Technical requirements

You can download the supporting files for this chapter from GitHub at `https://github.com/PacktPublishing/Inkscape-by-Example/tree/main/Chapter02`

Using Inkscape for logo design

As logo design is an important job for anyone working in the creative field, it is only natural to expect Inkscape to be up to the task as well. But what are the criteria of software that you can use to create awesome logos? Let me answer that question from another angle: what makes a good logo?

First of all, a good logo has to be **versatile**. Technically, this means it has to be vector-based. Vector files are scalable, editable, and portable. They allow us to transform the final logo into any color or size to use it on (almost) any surface. A vector file can be exported into the most common standard image file formats. (You will learn more about formats at the end of this chapter and even more in *Chapter 7, Combine Inkscape and Other (Free) Programs in the Design Workflow*)

Second, a good logo is **simple**. Most people think it takes no effort to create a simple design. But to create a simple but creative logo, you need to work a lot. Simple means easy to remember and redraw with a few elements, but this simplicity must be achieved first.

And finally, a good logo needs to be **creative!** A creative logo stands out; it is a unique visual image. A creative logo is built on a lot of thinking, sketching, and trial and error.

Being a vector graphical program, Inkscape has all the tools you can use to create a versatile logo. It can export to different formats, as well as take full advantage of vector shape and color editing.

Starting the logo design process with the usual first steps

Apart from the usage of Inkscape, in the following paragraphs, I will also share my workflow to design clever logos.

You should start designing a logo with Inkscape the same way you would start using any other graphical program. And all the designers agree that the process does not start with opening that program itself! You have some work to do before drawing anything.

We usually start every design process with the same three steps: gather information, think, and sketch.

Reading the brief and doing your homework

The very first step is learning more about what you have to design. In this case, we have it easy since I will provide the information needed.

Logo scenario

In this chapter, our task will be to create a logo for a very successful – yet fictional – company called CloudUsers. They are a small company offering cloud-based data solutions to its clients. They offer services such as hosting, data recovery, migration, virtual machines, database services, and so on. Their main values are safety, stability, and speed; this is what you have to emphasize in the visuals.

Reading the brief covers only part of the story – you usually need to do research and learn more about the company, the market, and the competitors. In this case, if you did a bit of research, and looked at the logos of other companies with similar profiles, you would see that a simplified cloud is a common symbol for cloud-based companies. Although the servers are not literally in the clouds, this is an easily recognizable image.

Thinking and sketching

As the common cumulus cloud became an emblem for cloud computing and related services, we should include it in our logo design.

But what else is needed? What would make this logo blend into the tech field, yet stand out from the crowd? This is where the initial sketches come in.

Sketching with a pencil on paper helps you decide what you need to draw. It is the method of grabbing the ideas swirling in your head and putting them down on the paper in front of you. In my opinion, as a designer, sketches are for getting your ideas straight, and they do not need to be final by any means.

A cloud with rain or the Sun is a suitable icon for a weather app, but an IT company needs to have tech elements too. And as the company name suggests, this logo needs to have users or people in it:

> **Tips about sketching**
>
> During the sketching period, I usually write out the name of the company several times, to look for typographical opportunities to add an extra layer of meaning through the shape of the letters. I also draw elements from the name and the profession of the company. Remember: sketching is not an art form. It is about drawing a ton of bad drawings and looking for visual solutions to your design problem!

Figure 2.1 – A few of my initial sketches for the CloudUsers logo

Importing your sketch into Inkscape

Now that the initial sketches are done, it is time to open Inkscape.

As I mentioned previously, sketches are never meant to be finished designs for me. This means I am glancing at my sketches most of the time while working in Inkscape, but I am not using them strictly as a base to trace over.

Other designers use their sketches differently and like to work out their ideas fully on paper first, and later follow these sketches precisely in the digital design app they use. It is all a matter of preference. But even if you only use your sketches as rough references, it is a good idea to import them into Inkscape. That way, it will be easier to find the final form that matches your ideas.

Scan your sketches or take a picture of them when you plan not to follow them precisely. Since this project was created as a step-by-step tutorial, you will learn by following my logo design in Inkscape, so you do not need to import sketches at this point. If you want to use my sketches, you can download them from here: `https://istvanszep.com/inkscapeexample/logo-sketches.jpg`.

Click on **File** | **Import** or press *Ctrl + I* and import the picture into Inkscape. Because the bitmap image will not be used in the final form of the design (that is, it will be a vector logo, not a photomanipulation), choose the **Link** option under **Image Import Type** from the import popup window.

Linking your image instead of embedding it will keep your file size from bloating performance issues. Again, since you do not want to include the photo in the final design, high image quality is not a must. So, just leave the other import settings as-is:

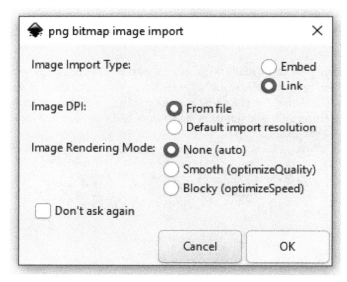

Figure 2.2 – Linking your sketches while importing them

This concludes the very first steps of the logo design process. With or without Inkscape, you always need to work out what you need to design. Before starting up the program, always think and sketch! Then, import your sketch into Inkscape and get ready to take the next step!

Designing your logo in Inkscape

Shape, color, and text are the absolute keys to logo design. In this section, you will learn a bit about the theory behind them, but we will focus on how to implement them in Inkscape! First, you will build the shape of your logo, then apply colors to your design. Finally, you will add text to finish the logo! Apart from these key steps, you will also learn a very efficient method for decision-making in Inkscape!

Step 1 – turning the sketch into a vector

Now that we have some ideas in the form of sketches, it is time to turn them into proper vector shapes. The cleaner the shapes of a logo, the better it will be. This creates simplicity and helps people understand the image. Clean shapes are also based on geometry. The simplest way to achieve pleasing geometrical proportions is to recognize the basic building block of your logo and use that in Inkscape to design it.

I picked the most creative from the sketches – the one built up from simple user icons while still resembling a cloud. Now, we have to recreate the winning sketch in Inkscape.

Take a look at our sketch and deconstruct it. It was purposefully built up from the basic shape of those well-known user icons we see everywhere. That *user* shape can be rebuilt using simple rounded rectangles and circles. You will have to create this shape first:

1. Draw a tall rectangle and use the small circle handles at the corner to round it. Feel free to use any color at this stage; we will apply colors and gradients to the logo later on. If you want to draw over the sketch to check the proportions, set the object's opacity to 50% for transparency. Otherwise, just keep it at 100% and draw your shapes next to the sketch:

Figure 2.3 – The winning sketch and the basic shape of the logo design

2. Draw another rectangle over the first one horizontally. Select both of them and choose **Path** | **Difference** by pressing *Ctrl + -* to cut the bottom of our first rounded rectangle off. This way, you will create the body of our user figure.

3. Now, use the **Circle** tool and draw a head over this torso while holding *Ctrl* to create a perfect circle. In this design, keep the proportions shown in *Figure 2.4*.

4. Set the size of the head to half of the width of the torso. Then, use the **Path editor** tool and adjust the height of the body to be 3.5 times the size of the circle. From now on, for this logo, we will use the size of the circle as one head unit, to keep the measurements and proportions right.

Tip

When editing the body shape, only move the two bottom nodes up or down with the **Path editor** tool. Do not use the **Transform** tool, as this will stretch it and change the proportion of the whole shape. This is an important tip to remember when you create a design based on different variations of the same object.

5. Press *Shift + Ctrl + A* to open the **Align and Distribute** window. Select both the head and the body, and center them on the vertical axis by clicking on the corresponding icon in the **Align and Distribute** dialog window. The tooltip of the icon says **Center on vertical axis**. Set the distance between the head and the body to be about a third of the diameter of the head circle. With that, you have finished creating the first user shape for your logo design:

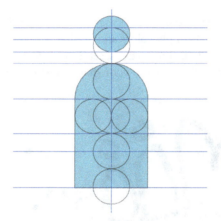

Figure 2.4 – Proportions of the two elements in the user shape

Now, create the second user shape. According to our sketch, this has to be shorter than the other so that it can build up the cloud shape illusion. Plus, it will look like a sitting figure compared to the taller one:

6. Select both the body and head of the first figure and group them by pressing *Ctrl + G*. Now, duplicate them by pressing *Ctrl + D* and put them next to the first one.

7. Set the distance between the two users to half of the distance between the head and the torso. If the body height of the taller figure was 3.5 head units, set the body height of this one to 2.5 head units using the **Path editor** tool!

8. Select the two users and put them on the same baseline via the **Align bottom edges** icon in the **Align and Distribute** dialog. Notice how the measurements are lining up: the shorter figure reaches the chin of the higher one. That is the point of creating geometry and structure!

Figure 2.5 – The two users next to each other at this stage

Now that the user part of our logo is laid down, it's time to focus on the *cloud* side. As I mentioned previously, your logo design is based around the same half-rounded rectangle shapes. You only need three more shapes to finish this simple logo design:

9. Duplicate the taller user shape and pull this duplicate to the left.

10. Set the distance between this and the original to the same as the distance you used before. This way, the *user* part and the *cloud* part will be a bit separated.

11. The two parts of the duplicated user – the body and the circle – are grouped now, so ungroup them by right-clicking on this group and selecting **Ungroup**, or pressing *Shift + Ctrl + G.*

12. Now, delete the circle here since you do not need that to create the cloud shape. What you need to do is make this remaining shape taller, to signal the peak of the cloud. Using the **Path** tool, select the top nodes of this shape and lift them until the height of the shape is 5.5 times the height of the head unit. This height difference between the cloud and user elements will create a geometric triangle shape with harmonic proportions.

13. To create the second cloud shape, just duplicate the one you just made, and pull the duplicate to the left. You want to keep the users separated and recognizable as independent human forms but want to show the cloud parts as one piece. The reason is the same: you need the cloud to be one visual mass and recognizable as a cloud. So, this time, do not set the duplicated shape apart from the other shape! Let them overlap halfway, with the edge of the original cloud shape running at the center line of the duplicate.

14. Now, using the **Path editor** tool again, grab the top nodes of the new shape and set the height of it to four head units. Check *Figure 2.6* for how it all should look now:

Figure 2.6 – The cloud part of the logo is overlapping; the users stand separate

15. Adding the last part of the cloud is tricky, but you will use the same building blocks you just used earlier. Duplicate the last shape you created and rotate it 90° counterclockwise so that it is laying on its side. The easiest way to do this is to hold the *Ctrl* key while rotating it.

16. Select this last shape and the first cloud shape and use the **Align and Distribute** dialog to align the edges at the bottom and on the right-hand side. If you did everything correctly, you will have an almost rounded cloud shape ready! Well, almost, because the cloud illusion is not there yet:

Figure 2.7 – All the parts are ready, but the cloud illusion is not there yet

The piece that is breaking the illusion of the cloud is the user shape on the right edge. It has a straight and edgy bottom-right corner, not a rounded one like the cloud has on the left-hand side. By applying a curve here, you will fix the cloud illusion! Since the cloud curve on the left is the same shape as all the other ones in your logo, you can use that to shape your curve perfectly.

17. Double-click the user shape on the right to go into the group.

18. Now, select and duplicate the torso part of it and rotate it 90° clockwise (hold *Ctrl* for a perfectly horizontal result).

    Selecting the original and the duplicated shapes aligns their bottom edges and their right-hand sides via the **Align and Distribute** dialog! The reason you had to do this was to use this shape as a blueprint to shape that straight edge into a perfect curve.

19. From the **View** menu, from the **Display** modes, select **Outline or Outline overlay**.

20. Select the original shape – the one with the corner – and add new nodes with the **Path** editor. Simply look where the path of your blueprint object starts, and double-click on the original path at the same two places:

Figure 2.8 – Curving the corner of this shape in Outline Overlay display mode

---

Tip

To measure and build your shapes here, I suggest using the Outline view, to see where the edges of your shapes are overlapping. Even better, Inkscape 1.0 introduced the Outline Overlay, which allows you to see not just the outlines but the fill colors of your vector shapes, making them easier to identify! *Figure 2.8* shows the Outline overlay in work!

---

21. Now, simply delete that one stray node in the bottom-right corner and shape the resulting curve to perfectly fit the blueprint path! When you are done, just delete the blueprint object since you do not need it anymore.

22. Congratulations – the shapes of your logo are now ready! It has a more friendly and cloud-like appearance than before:

Figure 2.9 – The finished shape of the CloudUsers logo

Duplicating during the creative process

Although this chapter is a straightforward project for recreating one particular logo design for practice, it is worth mentioning at this point that there is another way Inkscape can serve you during the logo design process. The program lets you explore and create, giving you a safe space to develop your ideas further! It is a clean transition from experimenting with ideas on paper to experimenting with a more flexible digital canvas.

Any time you hit a decision point in your design work, where you think about changing the color, using a different font, changing a shape, and so on, instead of applying the changes straight on, create a duplication first!

We duplicate for two reasons: creativity and effectiveness. Stop wasting time imagining how the logo would look with another color! Duplicate it and test the two versions side by side in the same file! Then, move on to the version that works better for the project and keep working on that.

But keep the previous version as well – this way, if a change does not work out, you can easily find any previous versions and start branching out from that again. This is an important part of my logo design workflow.

I use duplication to create a whole tree of logos branching out into different versions of the same logo. This is a rule you should try to keep in your design process, and never apply hard changes to the same design.

How to implement this method

If you need a new version, just select all the elements of your current logo, and hit *Ctrl + D* to duplicate it. Then, holding the *Ctrl* key, move the duplicated version forward horizontally or vertically. This way, you will create an easily accessible tree of different versions. Here is a real-life example of how my average Inkscape logo design file looks while testing different logo ideas for one of my clients, *AI media research*.

The flexibility of the Inkscape interface allowed me to be flexible as well. You can experiment almost as freely as on paper – and sometimes even more. You can compare all the versions in a moment:

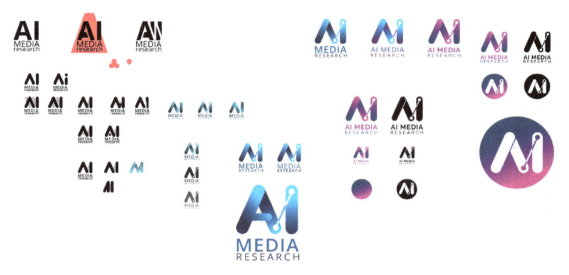

Figure 2.10 – This is how my average Inkscape logo design file looks while testing different logo ideas

When to use this method

As mentioned previously, whenever you need to make a decision. But more precisely, duplication is very useful, especially in the following cases:

- When you want to make changes to the shape of any logo element, such as its size, shape, position, and so on
- When you are choosing a color for your logo
- When you try different fonts or text placement in your logo
- When you merge elements in your logo design
- When you create different sizes and final versions of your logo

Step 2 – applying color and gradients to your logo

Everyone with a bit of interest in design has heard about color theory and the psychology of colors in logo design. The basic rule of simplicity also applies here: keep your color palette relevant but limited. Try to express the mood of your logo with as few colors as possible. Limiting the number of colors is also crucial for the usage of your logo design.

The fewer the colors, the easier it is to recognize the logo, but it also makes it easier to replicate it on any surface.

Colors in logo design are your ultimate device to express emotions and create a bond with the viewer. That is why it is important to research color samples for your project and look for different color moods to use in your logo before choosing the final colors.

If you find a color that you like, copy the three RGB values or the hexadecimal code of it into Inkscape, like you would do using any graphical program. A hexadecimal code is usually a # followed by six digits. But in Inkscape, you will see eight digits. That is because Inkscape is using RGBA.

RGBA stands for the RGB colors plus the Alpha value, which is the transparency of the given color in hexadecimal code. This is more information that can be handled with six digits, hence why you need eight. You can find this code in the **Fill and Stroke** dialog window, and it is the easiest way to copy colors from one place to the other.

Whenever you find a hexadecimal color code of six digits, just copy it into its place in Inkscape; the program will automatically add FF to the end of it to show it is a non-transparent color with a maximum Alpha value.

In this case, the cloud reminds us of the color of the sky, so blue is a good first choice. Also, most tech companies prefer a blue logo, since it stands for calmness, trust, and knowledge. To break the blue up a bit, you can also add another similar color and apply a gradient.

I picked **turquoise** since it is fresh and optimistic, and I aim to lighten up the tech part of the cloud for the human side of the logo. It is also good to have another color to work with later on while developing the rest of the visual identity. Gradients are trendy (yet again) and even a subtle gradient created from two fairly similar colors gives you a slight 3D effect.

The cloud logo is built up from multiple shapes, which gives you two options to apply a gradient. You can either select all the objects and merge them into one by selecting **Path | Union** before applying the gradient, or simply select the objects and group them before creating your gradient.

If you choose to merge the shapes, it is wise to duplicate your logo first, to keep the version with the separated elements as well, so that you can come back to it if you need to change something later.

Select the group of merged objects and apply a diagonal gradient of blue (**RGBA**: #0088aaff) and turquoise (**RGBA**: #37c8abff). Position the gradient handles inside the borders of the logo. This way, you will create a clean gradient with both colors visible at the edge, as shown in the following figure:

Figure 2.11 – Applying the gradient to the cloud logo diagonally

With that, you have added the primary colors. Now, let's move on to adding text to your logo. You will color the text as well, and you will learn about creating the different color variants later in this chapter.

Step 3 – adding text to your logo and modifying letters for more personality

Every company or organization has a distinctive name, so, naturally, every logo needs a version containing that name.

As a designer, you have three choices when working with text in logo design:

- First, you can create a unique typeface for your logo. This can be achieved with Inkscape using the **Path** tools, as we did with the logo design. Creating this type of *word art* logo takes a lot of work and practice though, and most of the time, it only works well if the text in the logo is a short word or abbreviation.

- On the other hand, you can pick and use an existing font that matches the feeling of your logo design, and helps you convey the message better. The duplicating method mentioned previously works great with this in Inkscape, making it very easy to compare the different font variations.

- Your third choice lies in between those two: picking an existing font and tailoring it to your needs so that it matches your logo design better while you're still maintaining a unique look.

We will use the third method with our CloudUsers logo design. You will modify an existing font so that it matches the shape of the cloud emblem we created earlier.

Because it is a small and friendly company, with other small and friendly companies as their target group, we will use a fresh and simple font for our text. The font family I chose is **Montserrat**, a **sans-serif** type font with several variations, including a bold style that fits the image of the CloudUsers logo almost perfectly.

Your task is to modify the shape of the letters a bit to turn that *almost perfect* into a perfect fit:

1. First, type CLOUDUSERS in capital letters under the cloud emblem you created.

2. Place the text under the emblem and align it roughly to the vertical center of it. You will arrange it and find its final position later after you have changed the shape of the text.

3. Select the text and set the font to **Montserrat** and the font weight to **Bold**. Feel free to use any color at this point as we will concentrate on the shapes for now.

Looking at the logo and the text in this state, you will notice that the letter *C* at the front is a bit off. Why? Do you remember the shape you created to build our cloud? That half-rounded rectangle is present in many letters in this clean and modern font.

This is yet another reason to use **Montserrat** as the typeface for this design: the letters *U*, *D*, and the curve of the letter *R* are all based on the same shape you based your logo on! But that letter *C*, although very nicely designed, could be forced to fit the other letters in the word. This is the letter you will modify, or, rather, recreate in a new form:

Figure 2.12 – This typeface fits the cloud image, but the C needs to be reshaped

To be able to modify the shape of the first letter, you need to turn the whole text into a path:

1. Select the text object and turn it into a path by selecting **Path | Object to Path** from the menu, or pressing *Shift + Ctrl + C*.

2. When you turn a piece of text into an editable path, all the letters inside the text object will turn into individual shapes, held together in a group. It is up to you to break up the group (by right-clicking it and choosing **Ungroup** or pressing *Shift + Ctrl + G*) or keep it and work inside it. You can enter a group by double-clicking it, or by pressing *Ctrl* and clicking on the object you need to modify.

Tip

Typography is the art of creating and arranging text. There is a whole list of visual rules a typo graphist has to follow while designing letters and typefaces. Only break these rules if it's necessary to your design and try to keep the visuals of the text consistent. Modifying one of the original letters is a good way to keep your design inside these set rules, while other actions (for example, distorting a whole word by skewing it) are not.

3. Select the original letter *C* and delete it. You could make your own version by using the shape on the left-hand side of the cloud, but it is easier to create something similar while reusing another letter already present. This way, you can also keep the thickness of the strokes and the shape of the letters in the text the same.

4. Your goal is to recreate the shape on the left-hand side of the cloud. The best letter to use here is the *D* in *CLOUD*. Click on the letter *D* and duplicate it by pressing *Ctrl + D*.

5. Then, move it in the place of the original *C* while holding *Ctrl*.

6. Push *H* to flip it horizontally, then cut off the straight line in the mirrored letter *D* to open it up.

7. Draw a rectangle over that vertical stem, select both the letter *D* and the rectangle, and select **Path | Difference**, or press *Ctrl + -*. This will cut the letter *D* into an almost perfect open *C* shape:

Figure 2.13 – Using a rectangle to cut a rotated letter D into a C

8. After cutting off the right-hand side of the letter *D*, you will notice that the letter *C* you created looks shorter than the rest of the letters in the row and seems out of the picture. But since you are using this technique to keep your letters consistent, you need to fix this by selecting the nodes on the right of the letter *C* and moving them to the right until the letter is wide enough:

Figure 2.14 – Moving these nodes created a wider look for the letter C

The shape of your text is now much better, fitting the cloud shape on the left.

9. To position your text, select the group containing the letters, or if you ungrouped the word earlier, group them again.

10. Then, select the cloud emblem as well and center them on the vertical axis using the **Arrange and Distribute** dialog.

The only thing you need to do is apply the colors so that they match the cloud emblem. You used two colors as a diagonal gradient on the cloud emblem, and now you need to apply a color matching that. Instead of using the same gradient for the text, let's apply another new color. But how do you decide on a color that is matching?

One of the easiest solutions is using the average color from the gradient present in the logo.

11. Select the text, then use the **Color picker** tool on the emblem to get the color data from there.

12. If you click and hold while using the **Color picker** tool, a circle will appear around your cursor, and Inkscape will calculate the average colors of the area inside the circle. This is a great tool to use here since it will produce a new color, but a color that is already there and is in harmony with your design. Feel free to use the color I picked for the text – its RGBA code is `#18b7c2ff`:

Figure 2.15 – The finished logo design with text and colors

> **Tip**
> Using the **Color picker** tool to extract the average color from any area works great in Inkscape. The only thing to note is to be careful about what is in that color selection circle since Inkscape gathers all the color information, including opacity and alpha values!

The main form of the CloudUsers logo is completed with a friendly geometrical shape and custom text, all presented in a fresh color. But as you know, the logo design process does not stop here. After the initial creative part has been completed, you need to work more to make your logo ready to use!

Creating logo variations for different usages

Thinking forward is the key when finalizing the logo you just designed. Every logo will be used on different surfaces and platforms, and as a designer, it is your task to prepare for that.

You should always keep an editable version of your logo design with all the graphical elements still separated. This ensures that if a change or rework is needed, you can fix things easily and you don't have to recreate anything that is lost.

Now, this being said, every other variation – the ones you hand over to the client – should be as simple as possible! Do not give out a finished logo as a group of objects; instead, merge all the shapes that have the same color, or all the objects that are possible to be merged.

Turn the text into a path to avoid lost fonts, and do the same with strokes as well. This way, you lower the chance of possible errors, and everything will stay in the same position and proportion as intended while designed. The CloudUsers design in particular does not have strokes, and the text is already a path since you modified the font; these are just general guidelines for logo design in a vector.

The following list will show you the most used variations and how to create them. You can see these versions in the same order in *Figure 2.16* going from left to right:

- **Primal logo version with gradient and text**: This is the main version, which shows the logo in its full glory. For this version, do the following:

    I.   Select the group of the cloud logo and ungroup it.

    II.   Then, use **Path | Union** and merge all the objects into one. It will keep the same gradient, but the elements will be harder to move individually.

    III. After this, do the same with the text group.

    IV.   Ungroup it, select all the letter shapes, and weld them together with **Path | Union**. You have two shapes now, one is the text, and the other is the iconic part of the logo. Keep them like this; do not merge them since they have different colors.

    V.   Select both of them and group them.

- **Flat color version with text**: Gradients are trendy, but they are not for every use. You also have to pick the main color that will be used when only one color is available. For the cloud logo, it should be the blue you picked for the text:

    I.   Duplicate the primal logo version and ungroup it.

    II.   Use **Path | Union** and merge the text and the cloud emblem.

    III. You now have a single shape; select it and set its color to the blue you picked previously.

- **White version**: Duplicate the flat color version of the logo and set the color to white. This version will be the one that can be used on colorful backgrounds, applications, or as a watermark on marketing materials, so it is great to have it ready at hand.

- **Black version**: Again, duplicate the white or the other flat version, and set the color of this new logo to black. This can be used for screen-printing, engraving, laser cutting, and on any other surfaces where color is not needed, or not possible.

- **Gradient version without text**: Simply duplicate the primal logo version and delete the text part from the group. This will be the version for occasions where the company name is not needed. For example, you can use this as a brand profile picture on Facebook, since the name will be almost unreadably small on the image itself, and it will be written out automatically next to the profile picture.

- **Gradient version with different text placement**: Since the CloudUsers emblem with the name under it generates a rather tall logo, we should make a variation with a more horizontal layout. This version will be great for website navigation headers, banners, or document letterheads:

    I.   Duplicate the primal version and ungroup it.

    II.   Now, place the text shape next to the cloud shape and scale it up using the **Transform** tool.

- **Flat color versions with different text placement**:

    I.   Duplicate the previous logo version you just created.

    II.   Ungroup the text and image elements and merge them into one shape with **Path | Union**.

    III.   Then, apply the blue color to it to create a one-color version from this text placement as well.

    You can also create a white and a black version from this layout:

Figure 2.16 – All the different logo variations you should prepare

So far, we have looked into the different logo versions you need to create. These are very useful to know, but they are only visually different from each other. Now, you have to combine them with the appropriate file formats to export to.

Exporting and formats

Most of the logo variations should be exported in all of the following formats to be used smoothly. Of course, there are some exceptions, but it all comes down to one thing: usage. Generally, vector files are good for printing, while a small bitmap image is all you need on a website. These are the go-to formats when exporting the final logo design from Inkscape.

SVG

SVG stands for **Simple Vector Graphics**, and this is the most familiar to any Inkscape user as this is the native format of the program. But as a W3C standard format, SVG is also the best vector format used on the web! It is small and scalable, so it's perfect for website headers and icons.

Inkscape can save to different types of SVG formats! While the Inkscape SVG has all the descriptions and metadata the program is using, when saving for web usage, it is better to choose plain SVG or optimized SVG. This will make the file size even smaller and the SVG graphics will be ready to be animated or scaled on the web.

Adobe Illustrator can also open SVG files created in Inkscape, so this is a good format to send to the client when they want to edit the file but do not use Inkscape.

EPS

Encapsulated Postscript files or **EPS** are versatile and considered the standard format for printing vector graphics. Whenever your logo is screen printed on a T-shirt or used in a brochure, EPS is a good file format to choose.

PDF

As the true **Portable Document Format**, **PDF** can be viewed on any platform, can be used in printing, and is harder to edit. In my interpretation, this means it is harder to create any mistakes after you have handed over the file to the customer.

PDF is also the best solution if you need to export in CMYK color mode. You will learn more about this in *Chapter 8*, *Pro Tips and Tricks for Inkscapers*.

PNG

Why is PNG on this list? PNG is small, and as a bitmap, it can be used on the web, on Facebook, and on other platforms. It also has transparency, so it can be used on different backgrounds – that is, as a watermark on a photo. This is the reason why a PNG logo, although not as scalable and editable as vector formats, is a must to send to your clients as an easy-to-use solution.

Summary

In this project, you drew the shape of your logo by using geometric principles and kept things in order. Then, you colored the logo and added text to it before learning how to create all the different logo variations and formats you might need during your work.

What you have now is a versatile logo that is a great base for any follow-up design project. Don't believe me? We'll put this theory to the test in the next few chapters!

3

Modular Icon Set Design with the Power of Vector

In the previous chapter, you designed a vector logo. When you design a logo, your aim is to create a single image of high quality. But when you design a set of icons, prepare yourself to create quality in quantity! In this chapter, you will create a modular set of nine icons for our project!

First, you will learn about icon set design and the different types of icons, then you will start by designing your first icon. Finally, you will move forward and create the whole set of icons in the most effective way possible, by cleverly implementing modular design, and reusing and redesigning your vector elements!

These are the main topics we will cover in this chapter:

- Learning about icons and icon styles
- Creating an icon is simple, but how about a whole set?
- Designing the first icon and the building blocks of the set
- Building the icon set using modular design
- Exporting your icons

Learning about icons and icon styles

Icons need to be very **descriptive**; they should be a simple representation of their original function. You don't need to reinvent anything, just do your research, and pick the metaphor you need. Be creative but aim for the easiest recognizable form.

After all, you want the user to recognize what is pictured by the icon. Safety? A lock, a key, or a shield are all safe bets. Speed? A speedometer or a flash. Ideas? A lightbulb or a brain. These are all common symbols you may use in your icons to make them instantly familiar to users.

Having a clean and recognizable form also helps with **scaling**. We usually design icons to be pixel-perfect in a small size with given measurements in mind. The times of using only 16x16 pixel icons are long gone; now, end users are used to seeing icons anywhere from 24x24 up to 180x180 pixels.

When you finalize an icon, it still has to cater to be used in small sizes as well. Keep the shapes of the icon clean and simple, so the unnecessary little parts do not add any noise to your design.

Before starting to create the icons, you need to choose a style you want or need to use for a project. In the following sections, we will categorize icon styles based on their visual complexity. This is not just about how the icons look, but also about how people tend to use them and how designers are building them up. Here are three examples of possible styles for vector-based icons:

- Line art icons
- Flat icons
- Illustrated icons

Line art icons

Line art icons tend to be minimal looking, with only a few details. They are usually designed using only strokes and no fill colors. This means the function of line art icons is clean and visible if the function itself is simple. A small icon for a *user*, *messages*, or *delete* is easy to create with a few lines, but when you need to picture something such as *client-side data deduplication*, you will need to add more and more lines and details to tell the story.

When creating line art icons in Inkscape, you can use regular objects to create geometrical order. This makes the icons more appealing, but only if you really focus on the rules of geometry.

Figure 3.1 – Reliability, safety, and great prices as a set of line art icons

Flat icons

Flat icons take details one step further than line art icons. Instead of using only strokes, designers of flat icons create simple flat shapes with color fills for better understanding. With no shading and a limited color palette, flat-style icons create a simple image that is easy to recognize. They are a great choice when you want to add more color to your icon set, and still keep things clean and simple:

Figure 3.2 – Simple, flat, medical icons about epidemiology

Illustrated icons

In this book, I call these **illustrated**, while others call them **decorative** icons, or even **3D** icons (compared to the 2D of the flat style). These types of icons have a lot of details – shading, gradients, highlights, and many colors and objects – so they are sometimes closer to an illustration than a simple icon.

But the function of an icon is there: they are an image that is representing a task or expression. We usually use these to get attention, tell a bit of a story, and highlight the most important tasks in an app or website.

Figure 3.3 – Illustrated cat icon set created by Erika Szep-Biro

Of course, the preceding icon styles are not set in stone. There are many more, and styles can mix and change as trends tend to shift. In general, the more complex an icon, the more of a story it can tell with all the added details and colors. Complexity is related to the visual hierarchy as well. The icon bearing the most important function is usually prominent and colorful, while others are simple shapes, such as that small **x** in the corner of a page.

These are the basic icon styles we need to create during our everyday work as designers. Now that you have learned about these common styles, it is time to move forward and learn about turning a single icon into an icon set!

Creating an icon is simple, but how about a whole set?

Creating a single icon is (usually) a basic task, but icons are almost never used alone. They come in a set, and creating a whole set of icons for the same project can be a real challenge!

Custom icon sets are usually created for previously designed brands to be used in app user interfaces or on websites. This means that, apart from the functionality of the icons, they also have to be in style with the brand. To create an icon set, the visual elements used in your icons need to be consistent. This will turn a bunch of random icons into one set.

The visual elements you can use to create a coherent set of icons

There are many tools in your designer toolbox. Here is a short list of the ones that can help you turn a group of icons into an icon set:

- **Color**: Use the brand colors or pick a matching color.
- **Size**: All the icons in the set should be created for the same size to help visual consistency and easier usage. This also means a consistent thickness of strokes and scaling of the elements of the icon.
- **Background shape**: Square, circle, rounded rectangle, hexagon, and so on. If you want to use a shape as the background of your icon, pick one and stick to it!
- **Basic shapes**: Curved lines or straights and edges? Flats or strokes? Follow the icon style you chose for the set, and more importantly, always keep in mind that a simple shape can change the expression of any image.
- **Repeating visual elements**: A line, a circle, a colored blob, and so on. Repeating the same small element can create a connection between the icons.

The solution – modular icons

A modular design is a clever solution; it means that the designer is working from prebuilt modules or building boxes. In your case – using Inkscape to create a set of icons – modularity is the solution for both quantity and consistency. You do not have to draw all the icons one by one but rather, combine and modify the same simple elements, creating results faster.

The modular process is straightforward. First, you need to choose a style using references and sketches. Then, you define your basic building blocks to use and design the first icon. And finally, you duplicate and modify those building blocks to create your icon set. Modularity is here to make your work easier, but it does not mean a repetitive, boring visual design.

Be creative while using your building blocks – that is one of the goals, after all. As you will see, you will reuse and repurpose elements as well as modify them to create the visual effects you need.

> **What you will create in this chapter**
>
> You will design nine icons for the CloudUsers brand we established in the previous chapter. The icons will be also used on the website you will design in *Chapter 6, Flexible Website Layout Design for Desktop and Mobile with Inkscape*!
>
> The icons are *Secure, Fast, User-friendly, Access control, Disaster data recovery, Cloud storage, Big data, Remote support,* and *Cloud automation* representations.

The modular design will make your journey for the best icon set faster, and moving forward, you will now take the first step: creating the first icon in the set.

Designing the first icon and the building blocks of the set

A set of icons needs to be visually coherent. Naturally, the first icon has to set styles and define the visual rules for the rest of them. When we talk about *rules*, we mean guidelines that help you work faster and easier. The rules are creative, but once you have created them, try to stick to them.

Experiment on the first icon but think in a set. Once you have figured out the rules by working on the first icon, keep the line thickness the same, use the same set of colors, and use the same building blocks. All this is to achieve a similar feel for all the icons in the set.

Remember the icon styles we learned about a few pages earlier? During this project, we will design all the icons of the set in **line art style**. This means the icons will consist of simple strokes without fill colors.

The flexibility of Inkscape will help you establish and maintain your own design rules throughout the process. In this case, for the sake of practice, the following guides and building blocks will be your icon design rules:

- **Use the cloud shape**: As all the icons are related to the cloud services, you can include the cloud shape for the logo in the icons as the one common element. The cloud shape will be the main visual element connecting the icons in the set.

  Using the cloud shape will also set the standard size of the icons in the set. This will be a great reference point to size and position every other element used in building up your icons, plus it is an interesting background shape. The cloud will provide a frame for your icons.

  Figure 3.4 shows some examples of the icons you will create. The cloud shape will be intentionally broken in different places, and the other elements will be added to explain the message of the icon.

- **Colors**: The colors of the logo are blue and turquoise; these are the colors you will use for the icons as well. The cloud will be blue (#00aad4), and the additional elements will be set to turquoise (#2ac1b5). This will add to the coherence of the icon set, and it will also train the viewer to recognize the repeating shape of the cloud as a standard and look at the different elements as clues about the functions we're trying to depict in the icon.

- **Stroke width and style**: This has to be consistent too. Do not use **fill color** and set **stroke width** to 6 px and **stroke style** to **Round Join** and **Round cap** on everything you create for these icons. Set it once for the first icon, and the settings will remain the same for the rest during this exercise.

- **Size of all the icons**: The size in a set has to be the same; in this case, you will create the cloud shape and fit all other elements in the same size. In the examples in this chapter, I used a cloud that was 102 pixels wide and 62 pixels high. Feel free to experiment with your own sizing!

Figure 3.4 – A few examples of the finished icons to show the visual rules in action

In the next section, you will design the cloud shape as the first building block for your icon set and get ready to design the first icon.

Creating the cloud shape

First, you need to define the shape of the cloud in the icon. It does not have to be the same as in the logo, but as mentioned in the preceding section, it is good if it has the same characteristics. In the logo, you only built one-half of the cloud; the other side was two user characters. This time, you have to create a whole cloud. To do that, follow these steps:

1. Open the file containing the cloud logo you created in *Chapter 2, Design a Clever Tech Logo with Inkscape*. Select and copy the user shape (both the head circle and that rounded rectangle you cut in half earlier) into a new .svg document.

2. Start creating the cloud shape by duplicating the user shape four times. Keep three of the shapes standing vertically next to each other, while rotating the one on the left and on the right side, as shown in *Figure 3.5*.

3. Set the height of the vertical parts as shown in *Step 5* and move them closer together. You should have three vertical shapes of different heights. Also, change the length of the horizontal elements if needed and move them closer together to create a closed cloud shape.

4. To keep the proportions and the regular shape of the elements, only change the length of the objects by moving the nodes on the flat part of the shape! Using the **Transform** tool to make them longer would ruin the proportions here.

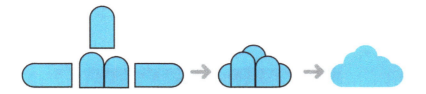

Figure 3.5 – Duplicate and rotate the user shape to create the form of the cloud icon

5. Now, select all five elements and merge them into one by selecting **Path | Union**. This will create one cloud shape. We want to create line art icons, so select the cloud, and add a stroke while removing the fill color. Follow *Figure 3.6* to get the proportions right. In my example, I set the width of the cloud to 102 px, the height to 62 px, and the thickness of the stroke to 6 px.

6. Set **Join** to **Round join**, and **Cap** to **Round cap** in the **Stroke style** dialog tab. This will also help when creating the icons later on when you will delete segments of the path of the cloud.

7. The outline color of the cloud will be the blue color of the logo (#00aad4):

Figure 3.6 – Cloud icon base with outline and no fill

This cloud is our main starting design element; you will create the remaining shapes later as we design icon after icon. Now that the cloud is ready, let's make the first icon from the set based on this shape!

Creating the Fast icon

Let's start with three icons explaining the advantages of the company's cloud-based solutions: *Fast*, *Secure*, and *User-friendly*. The icon used for *Fast* can be a speedometer, a lightning bolt, or lines implying fast movement. We will use the latter idea for this icon:

1. First, duplicate the cloud icon, which we will use as a base.

2. Then, modify the shape of the icon by removing a part of the stroke. Select and delete the node on the left side of the cloud curve. Now, select the remaining two nodes and remove the path connecting them using the **Delete segment ...** icon in the Tool control bar. The path is now open on the left:

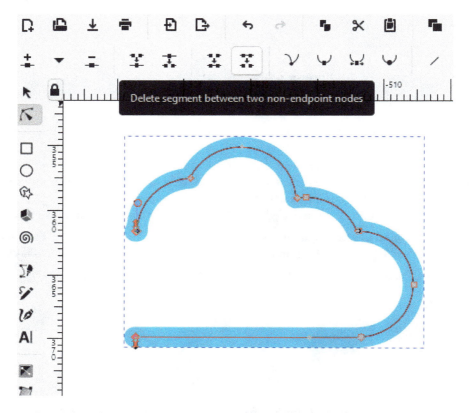

Figure 3.7 – The cloud shape is opened on the left with part of the path removed

> **Important notes about path editing**
>
> We will use the same method on almost all of the following icons! Duplicate the cloud base and open it up by removing parts of the path. This is a great method to work with strokes in Inkscape, and it is the base move for creating line art icons.
>
> You might think, why not simply use the **Difference** path operation to cut open the cloud shape? Because when you cut off a part of a path, and if the original path was closed, then it will remain closed with a continuous stroke. You would then have to delete the section of the path anyway, to open up the path.

3. Add the horizontal lines to create the speed effect. Draw a horizontal line with the **Bezier** tool (hold *Ctrl* while drawing to keep it totally horizontal) and set the color of it to the turquoise color (#2ac1b5) of the logo, and the stroke width should be the same as the stroke width of the cloud (in my example, 6 px).

4. Now, duplicate the line and position the two lines as shown in *Figure 3.8*.

5. Set the lines to different lengths to create the speed effect. The bottom line should be aligned with the baseline of the cloud, and the distance between the lines should be even. Also add another line attached to the cloud to make it more dynamic.

6. To do this, double-click on the path to add a node to it, and then move that new node a bit to the left. Take care, as the icon cannot be wider than the original cloud! Set the length of the lines accordingly. Here is my version of the final icon for the *Fast* cloud services:

Figure 3.8 – The final Fast cloud icon

This is your first icon in the row of nine. The more you create, the faster it will get because you will use the same cloud base for all of the icons; also, you will have more and more elements to reuse in the icons following the previous ones.

In this section, you learned how to open a path and add new segments to it to express motion. You also introduced another color we will use for this icon set.

Building the icon set using modular design

For the next part, you will create all eight remaining icons in a fast and effective way – instead of drawing everything from scratch, just think, build, and duplicate!

You will use the same working document for all the icons now. This is a very flexible solution because it allows you to duplicate and try different versions without opening and closing documents. You will be also able to see all the versions together and check for quality differences or possible issues by comparing them.

Working with these small line art icons is also easy on memory, so they can easily fit in a small .svg file without slowing down your computer. Let's continue with the second icon!

Designing the Secure icon

The sign for safety is usually a lock, a checkmark, or a protective shield. For this icon, we will use the padlock, as it is a simple but recognizable shape, and you can easily create it using one of our readymade elements:

1. First, copy the user shape into your working document (the half-rounded rectangle you used to build the cloud shape).

2. Add a turquoise stroke to it and remove the fill color.

3. Set the stroke to the same thickness as the cloud – in my example, 6 px. Now, put a copy of this shape aside, as we will use it later as a building block! The following figure lays out the steps you will take to design the padlock:

Figure 3.9 – The steps to create the padlock

4. Duplicate the user shape, select this duplicate, and rotate it 180 degrees. Position these two shapes under each other: the one above will form the shackle of the padlock, and the one under will be the body.

5. Now, you will select the shape on the top, and scale it smaller. But if you just scale it right now, the width of the stroke will scale proportionally with the shape.

   This is the default in Inkscape, but you do not want that while working on line art icons. What you want is control over the stroke thickness – in this case, you want to keep all the strokes the same 6 px width. So, before scaling the top object, turn off this icon in the Tool control bar when using the **Selection** tool:

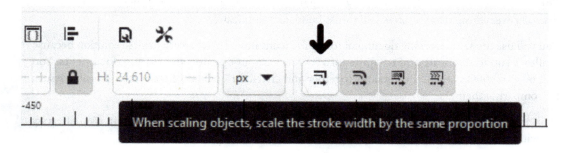

Figure 3.10 – Turn off this button to keep the stroke width the same while scaling

6. After scaling the top object, use the **Node editor** tool, select the two bottom nodes and delete the part of the path between them (using the same method you used earlier with opening the cloud shape).

7. Then, select one of the nodes at the opening and delete it to show that the lock is openable. Now it seems that the shackle is not rotated, just lifted a bit, and it makes it very recognizable as a padlock.

8. Now it is time to add a keyhole, and your padlock is done! It has to be as simple as possible, so draw a small vertical line in the middle of the padlock using the same turquoise color and 6 px stroke width.

9. Then, draw a slightly bigger circle on the top of the line, to make it more like an old-school keyhole.

10. Set the size of the circle to 8 px. (You will use this circle later too!) The padlock is now finished; select all the elements of it and group them.

11. To finish the icon, you have to add the cloud to the background. Create a copy of the cloud and put it behind the padlock. Match their position so that the top of the padlock is following the top curve of the cloud. Mind the height of the original cloud, and do not push the padlock higher than that!

12. To make the icon set more interesting, you will do the same thing as with the *Fast* icon and open the shape of the cloud a bit. To do this, create two nodes on the cloud on both sides of the padlock.

13. Select and delete any nodes between these two, then delete the part of the path between the two remaining nodes as you did earlier. This created the illusion that the shape of the lock completes the top of the cloud! You will play with this visual effect in the upcoming icon designs as well!

The *Secure* icon is now ready!

Figure 3.11 – Breaking the top of the cloud to make the final icon more interesting

---

The trick to working with paths

If you wouldn't add the two new nodes to *freeze* the path, deleting nodes of a path could change the curves connecting them to their neighboring nodes. This is a neat little trick if you want to modify a part of a path while keeping the other parts intact! You can spare a lot of headaches by saving your original paths like this, and not needing to reset them by hand when they accidentally change curve.

Designing the User-friendly icon

How can you communicate that software is easy to use? The simplest way is to show a happy and satisfied user. This is exactly what you will do for this simple icon too. You will draw a smiley face on the cloud. If this icon would stand alone, you could argue that it is not *techy* enough. But because it is part of an IT-related icon family, users will understand its meaning, and it will add a friendly vibe to the whole set.

To do that, follow these steps:

1. Begin making this icon by creating a smile on the face. You can do this in two ways (I will explain the difference between the two methods):

   The first method is using a shape you created earlier:

   I.   First, copy the user shape here and flip it 180 degrees. Select the two nodes on top of it and delete the flat part of the path between the nodes.

   II.   Now, add two new nodes closer to the center of the curve and delete all the nodes above them!

   This method will create a little smile that has the exact same curve, color, and stroke width as all the other shapes in the icon set! This is the way of thinking in modules and reusing elements in vector design. This is the method you can see in the next image.

   The second method is using the **Circle** tool:

   I.   Use the **Circle** tool and draw a small regular circle. Then, grab the white dot marker on the circle to create an arc of 120 degrees or smaller.

   II.   Rotate the arc if needed, so the two endpoints are on the same horizontal level. Now, this seems like the fast and easy way, but you have to scale the arc and set the stroke according to the other lines in the icon set. With the previous method, that is all set previously, plus the size will be given as well.

2. Add the eyes by using the dot you made for the keyhole on the lock. Duplicate it and position the two eyes wide enough above the smile to create a friendly face.

3. Create a copy of the cloud base here. This time you will not open the shape, but leave it as it is, forming a simple backdrop for the smile.

4. Finally, select the eyes and the smile, create a group of them with *Ctrl + G*, and position them in the middle of the cloud. See *Figure 3.12* for all the steps of this simple icon!

Figure 3.12 – All the steps to build the User-friendly cloud icon

Creating the Access control icon

The first three icons were more decorative and expressed the values of the company. Let's move on to icons depicting the concrete services of the CloudUsers company. This first one will be about managing access to different projects and networks on the cloud.

Access again could be a padlock, but to make it different, you will design a key and a keyhole to deliver the message.

To do that, follow these steps:

1. Let's start with the key, and draw a circle with the same turquoise color and the same thickness as all the lines before. Now add a horizontal line to it by copying one from the *Fast* icon. (This seems counter-intuitive, but sometimes, copying is faster than switching tools and setting up stroke properties.)

2. Now, duplicate this line and rotate it 90 degrees while holding *Ctrl* to create a vertical line. Make it shorter and position it closer to the tip of the key. Select this line and duplicate it, move it to the left a bit, and make it even shorter, to add another bit to the key. Select the circle and the three lines and group them. Your key is now ready!

Figure 3.13 – How to build a simple key icon

3. Now make a copy of the cloud base shape and position it behind the key. Also, copy the keyhole from the *Secure* icon you created earlier and put it on the cloud too.

4. Again, you will have to open up the cloud shape a bit to make it more interesting. Double-click to add new nodes to the path, delete the nodes between them, then delete the section of the path between the two nodes.

5. Now, position the key so it fills the position of the missing curve. Take care that the key is not sticking further out on the left than the original shape of the cloud. This way, all the final icons will keep the same size as the others. See *Figure 3.14* for more details.

6. This icon has to tell more of the network access management side of the story, so you need to add a part representing that. For this, draw a simple network node.

7. First, add a horizontal line under the cloud. Then, duplicate it, rotate it 90 degrees, and put it on the horizontal line to form an upside-down T shape.

8. Shorten the vertical line of the T, then select both lines, and make sure that the T is symmetrical by clicking **Center to vertical axis** in the **Align and Distribute** window. Group these two elements, then position this network line aligned to the keyhole in the cloud, and also align them to the base of the cloud.

> **Tip about deleting sections of a path**
>
> When you delete a section of a closed path – as in the cloud in this chapter – naturally, the path will keep the rest of its shape. Even when deleting multiple parts of a path, the nodes will still be kept together as the same path. They will be one path – you can scale them as one, move them as one, and so on. This is very useful a lot of times, but if you want to separate the parts of a path, you can also do that by clicking **Break Apart** in the **Path** dropdown menu.

9. You have to open the cloud path one more time (this time, at the bottom) so the network path has an opening. Double-click the cloud path on the two sides of the bar to add new nodes on the path of the cloud, then select these two nodes and delete the section between them. This way, the turquoise path will fill the gap at the bottom of the cloud. To maintain the spacing, adjust a bit the length of the vertical and horizontal paths if needed:

figure 3.14 – The last steps to create the Access control icon

10. Select all the elements of the icon and group them. The *Access control* icon is ready now!

Creating a small icon for Big data

Big data analytics is one of the key roles of cloud computing. The best metaphor for it could be a visual representation of a database, a diagram, or a mosaic of small dots or cubes. In this case, we will take a playful approach and combine the cloud shape with a rain-like mosaic of lines of code. It might sound like a lot, but it will be very simple really.

Follow these steps:

1. First, you need to create the data. The best way to do this is to create a few lines of zeros and ones in a minimalistic way that matches the style of the icon set we are working on.

2. Copy the keyhole shapes here: a small vertical line and a circle. Scale the circle to the same size as per the thickness of the line (in my example, 6 px). The line will be a perfect one, and the circle will make the zero.

3. Copy the cloud base here. First, we will use it as measurement only, while laying out the *data*. Duplicate a few circles and lines to start with and spread them out along the base of the cloud. As I wanted to keep the curve of the cloud on the sides, in the example, I used six elements in a row, both zeros and ones.

4. Now, select these six elements and distribute them evenly. To do so, use the button labeled **Make horizontal gaps between objects equal** in the **Distribute** section of the **Align and Distribute** window:

Figure 3.15 – Distributing the selected zeros and ones evenly

5. Select all the lines and circles in the row and group them. Now you have one line of data prepared. Duplicate it twice to create two more rows to have a mosaic. Arrange the two lines of data under the original one.

    As you can see now, the three rows are identical, and they look boring. Nothing about this communicates that it is a huge amount of data that needs to be understood.

6. To add a bit of randomness, double-click on a group to go into it, and reorder the elements. You can also duplicate and delete some zeros and ones, be playful, and create your own versions! Take care that the objects are in the same six positions as before so there are no horizontal differences.

7. Repeat this with another group too. Now you have three different-looking rows. Select them and group them together.

8. The last step is to add the cloud, and open it up – this time, at the bottom. Add two new nodes to the base of the cloud and remove the segment between them.

9. Align the cloud to the data such that the base of the third row of data is in line with the base of the cloud. This will move the first row closer to the edge of the cloud on the top. Take care that the first and last elements in that row are circles, so they are not touching the cloud line!

10. Now, the *Big data* icon is ready!

Figure 3.16 – All the steps to create the Big data icon

Creating the icon for Cloud storage

This one will be an easier icon to make. Cloud storage means accessible user data in the cloud, so you will create two simple arrows to stand for uploading and downloading data. You need to create the arrows first. There are many ways you can get this done: form the arrow from a triangle, draw it using the **Bezier** tool, or build it from simple lines.

This time, we will stick to the most efficient method and build the arrow from lines, so it is matching the existing shapes and styles in the icon set:

1. First, copy a single turquoise line from one of the previously built icons. If the line is not horizontal yet, rotate it now by holding *Ctrl*.

2. Duplicate this horizontal line and make it shorter by moving one of the end nodes while holding *Ctrl*.

3. Then, rotate this line while holding both *Shift + Ctrl*. Holding *Ctrl* will help you achieve a perfect 45-degree angle, while holding *Shift* will make the object rotate around the opposite corner of its selection box, not around the center as usual. (This is the fastest way for off-center rotations without manually repositioning the center of the rotation mark.)

4. Position the angled line to the end of the horizontal line to form a point. Now, duplicate the angled line and press *V* to flip it vertically. Position this new line to the other two lines to form the tip of the arrow. Select all three lines and group them.

5. Make a copy of the base cloud shape here. Take your arrow and rotate it 90 degrees by holding *Ctrl*. Duplicate the arrow, and flip it vertically by pressing *V*.

6. Now, position the two arrows on the cloud, as shown in the next figure. Align the down-arrow so it touches the line of the cloud at the corner. Align the up-arrow to the base of the cloud.

7. Create two openings on the cloud shape, one around each of the arrows. Use the known method: double-click the path to add new nodes, then delete the section between the nodes.

8. Select the arrows and the cloud and group them. You have finished the *Cloud storage* icon!

Figure 3.17 – The steps to create the arrows and the Cloud storage icon

Cloud automation icon design process

After a short and easy design, let's create something more complex. Although cloud-based automation is more about software solutions than hardware, automation is still best pictured with robots, factory assembly lines, or simple gears. For this icon, you will draw a set of gears in Inkscape. Apart from being easily recognizable, a gear is also the general icon for settings in an app, so it is always good to know how to draw them.

This icon needs a set of gears, not just one, as it has to be recognized for gears working together to create automation. So, you will have to design two gears at a minimum, but no more than three, to keep the design of the icon simple, and with it, the clean and simple appearance of the icon set:

1. Let's start with creating those gears! First, draw a small circle (the one I created is 22 px). Set the stroke width to 6 px, the stroke color to **Turquoise**, and turn off the fill color.

2. To create the teeth of the gear, add a short vertical line (with the **Bezier** tool) to the circle. Select both the line and the circle, and align them by clicking the **Center on vertical axis** icon in the **Align and Distribute** window.

> **Design note**
>
> In this design, we use lines with round caps, as all the shapes in the icon set are rounded and friendly. If this icon set would use edges, you could draw a square instead of the short line we used here.

3. Now you will create all the teeth along the wheel of the gear. To do this, you need to set the center of rotation of the tooth to match the center of the circle.

4. To find the exact center of the circle, select it first, so the arrows appear around the selection box, showing you where the center lines of the circle are.

5. Pull a horizontal guide from the top ruler and position it on the horizontal centerline of the circle. Now, because the tooth is aligned vertically to the circle, you only need to find its horizontal center. If you double-click on the tooth now, you can easily pull the rotation center (that small cross in the middle) down to the horizontal center line of the circle while holding the *Ctrl* key.

6. Now the center of rotation of the circle and the tooth are aligned. This means that, if you rotate the tooth, it will perfectly follow the shape of the circle, keeping an even distance from its center.

7. Duplicate the tooth and rotate it while holding *Ctrl*. Because you duplicated the first tooth of the gear, you don't have to align the center of rotation again, as it will stay in the same spot, right at the center of the circle. Duplicate it and rotate it again and again, until you have eight teeth at even angles around the circle. The first gear is ready now! Select all the parts of the gear and group them together.

(This method is also useful when you want to create similar images with shapes rotated around a circle, such as a simple clockface, or a flower with petals.)

Figure 3.18 – The steps to creating a simple gear symbol. The second step shows how to align the center of rotation to the circle

The *Cloud automation* icon needs three gears, with each looking different than the others. To keep the design together and make your job easy, this difference will come from size. To do this, follow these steps:

1. Duplicate the first gear and scale this duplicated gear to be a bit smaller. Remember the small icon in the top control bar you turned off a few pages earlier? That is still turned off, so Inkscape does *not* change the width of the strokes while scaling an object. The result of this is that when you scale the gear now, it will not just be a smaller version of the original, but it will have different proportions too! Feel free to experiment with the size!

2. To create the third gear, first, duplicate the original gear again. This time, you have to scale it a bit up to make it different. Notice how the teeth look seemingly longer and thinner this time because the stroke width is fixed. Again, feel free to experiment with the size! Keep this in mind and do not overdo it – keep the size bigger than the original, but only about 20%.

3. The third gear is bigger than the other two, but it also feels empty in the middle. Add a small circle in the center of it to fill the space there, and to make the gear more different from the other two!

4. The only task left is to position the gears on the cloud base to finalize the icon. Create a copy of the cloud here and move the three gears over it. This needs a bit of experimentation since your goal is to fit all the gears in the cloud. Try to align them to the curves of the cloud. Make the gears overlap the stroke of the cloud a bit. You can also rotate each gear too, so they are at different angles and the teeth of the gears are lining up in a natural way.

5. As the last move, open up the cloud shape again as you did with all the other icons in this set. Double-click the path of the cloud on two points before and after where it intersects with the first gear. Then, select both new nodes and delete the section of the path between them.

6. Repeat the same step for all the gears. When finished, select all the elements of the icon and group them. The *Cloud automation* icon is now complete!

Figure 3.19 – Creating two more gears and finishing the Cloud automation icon

Creating an icon for Remote support

Support is very important for any IT service or product. Usually, it is portrayed using a picture of someone answering questions, or a headset with a microphone. This time, you will reuse some of the shapes you made earlier and add a small character and a speech bubble to the cloud icon:

1. The small *User* icon is the easiest to make because its shapes are already created. So first, copy the user shape again.

2. Then, draw an 8 px tall turquoise circle above the *body* of the user shape to create the head. You can also just copy that 8 px circle you used before in some of the other icons of the set.

3. Select both the head and the body and center them vertically using the corresponding icon in the **Align and Distribute** window. Also, adjust the distance between the head and the body to about four pixels, or half of the height of the head.

   You can set this distance closer but not further, as proximity creates visual unity: the viewer will understand that the small dot and the body form a small figure together.

4. To make this character more dynamic, add a small arc to it, such as a waving arm. You can copy the arc created for the smile in the *User-friendly* icon or create a new one by using the same method. The aim is to keep the arc the same width, stroke, and mostly, size, as the top of the body of the character.

5. To create the speech bubble, first, draw a circle next to the body. Then reduce it to a 270 degrees arc by using the **Circle** tool or the **Path editor** tool and moving the white dots in position on the circle.

6. Now, draw a small angle to the speech bubble, to indicate that the speech is indeed coming from the character. Use the **Bezier** tool and draw a short horizontal line that is turning back up at a 45-degree angle. See this step and all the previous ones in the following figure:

Figure 3.20 – The parts to use to create the talking character, and the final Remote support icon

Create a copy of the cloud base and add the character and the speech bubble to it. Align the character to the baseline of the cloud, and the speech bubble to the curve of the cloud.

Add new nodes on the cloud path on both sides of the character and the bubble and delete the sections of the path, as shown in *Figure 3.20*. The *Remote support* icon is now finished!

Designing an icon for Disaster data recovery

The difference between this icon and the *Cloud storage* icon is that here, there is danger and urgency. You need to express that the service is made for accidental events such as fires, but it is also there to help the user and keep their data safe:

1. As a first step, let's see what you can repurpose from the elements you created earlier. You obviously need a copy of the cloud base, as with all of the previous icons. You also have a *User* icon and an arrow made previously, so create a copy of those as well!

2. This time, instead of straight arrows pointing up and down, you will create arrows that follow the curve of the cloud. You can achieve this in two different ways. You could create the curved lines manually using the **Ellipse** tool or the **Bezier** tool and then bend and position the lines carefully over the original cloud shape, or you can use the shapes of the path that are already there. Since this is a faster and more elegant solution, let's practice the latter one now!

3. Select the tip of the arrow, create two duplicates of it, and place them on the two sides of the cloud icon, as shown in the next figure:

Figure 3.21 – The steps to create the Disaster data recovery icon

4. Now you will turn the curves of the cloud into the shaft of the arrow. Using the well-known method, create a small gap above and under the arrows, as shown in the previous figure. Take care that you need to add four new nodes around both of the arrows, and delete only small segments, not the whole curve! You only want to create small gaps in the cloud path.

5. Now the arrows are curving and virtually separated from the rest of the cloud, but they are still part of the same path. You need to separate them so they can be colored differently. Select the cloud and click on **Break Apart** (*Shift + Ctrl + K*) in the **Path** menu. This will literally break the cloud path apart into four individual paths:

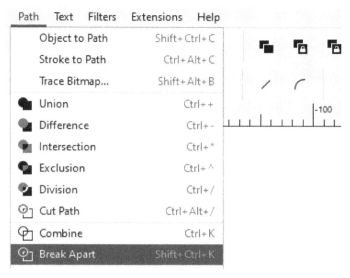

Figure 3.22 – Using Break Apart to separate all unconnected parts of a path

6. Select the two curves on the side, and color them the same turquoise you used in the previous icons. Now you have the cloud and the arrows on each side in a perfect shape.

7. Position the *User* icon in the cloud. In the *Remote support* icon, you added a small, curved arm to it, so it seems alive and helpful. In this case, your aim is to make it look panicky. So instead of a curved arm, draw a small, curved line diagonally over the shoulder of the small figure.

8. Duplicate the arm and mirror it horizontally by pressing *H*. Now, position this arm symmetrically on the other shoulder of the user.

9. To add more urgency to the icon, draw an exclamation mark. To do this, simply duplicate the head of the *User* icon. That is the same 8 px circle you used in many places before. Position it next to the user at the lower part and draw a vertical line above it with the **Bezier** tool to create the exclamation mark.

10. The only thing left is to position the user on the baseline of the cloud and cut an opening around it on the cloud path. Create a node on each side of the character, and simply delete the segment between the nodes.

11. Select all the elements and group them. The icon is now finished!

This was the final icon in the list of nine. If you created all nine icons in the chapter, you were able to practice a very important thing for a graphic designer: being resourceful! You can create one element and duplicate and modify it and have a set of icons, where each icon is unique but similar to the others:

Figure 3.23 – All nine icons created in this chapter

You created a whole set of icons in a short time, using duplication and the **Path editor** tool to open paths and modify them if needed. In the next section, you will learn about exporting your icons into two different formats.

Exporting your icons

For now, all the icons you designed while reading this chapter are in one file. This made it easier for you to duplicate and reuse the elements, and also experiment with them if needed. All of the icons in one .svg file are neat and easy to use during the design process, but of course, not suitable for the final goal of the icons.

In this book, the final goal is to use the icons as decorative and descriptive elements in the website design you will create in *Chapter 6, Flexible Website Layout Design for Desktop and Mobile with Inkscape*. To be precise, to use them in that project, you do not need to export the icons yet.

But we cannot wrap up a chapter about vector icon design without talking about exporting those icons in their final formats either. There are plenty of cases when a designer needs to export icons in a fast and reliable way, and there are easy solutions for these in Inkscape.

The two main formats to export your icons (or any images created in Inkscape) are PNG and SVG. Both have different properties, and knowing how to export them quickly and effectively is a useful addition to your designer toolkit.

Exporting PNG icons

PNG icons are still the go-to format for app user interface designers and website designers. As you know, PNGs are usually small files, plus they can handle transparency, which is a must for icons in a modern interface. If you want to use a background color change (for example, for interactions), PNG icons are great.

So how do we export PNG files for icons? If you only have one or two icons, just select the icon and export it manually via the **Export** window.

If you need to export more icons, I think it is easier to batch-export them. There is a tab for that in the **Export** window, where you can decide between exporting a single image or doing a batch export:

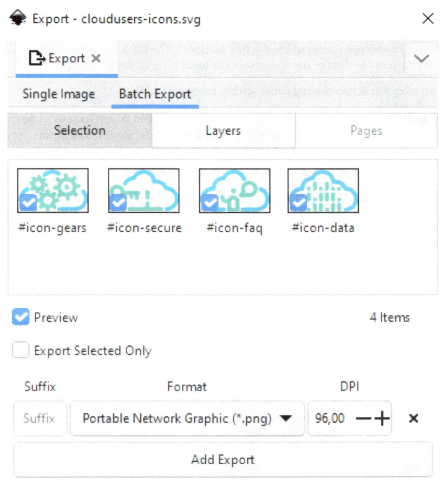

Figure 3.24 – Batch export can save you a lot of time

How to batch export PNG images

First, if you have not done that yet, select the elements for each icon and group them.

Then, select the nine groups, tick the aforementioned **Batch Export** checkbox, and hit the **Export** button. This will create nine PNG files of the icons you created. Now you can find the images and rename them in the folder you saved them.

The name of the exported files actually comes from the SVG structure of Inkscape. When you draw, every object is assigned a unique ID in that document. It is created from the type of object and a number. For example, an ID can start with `path` for lines and paths, `rect` for rectangles, or `g` for groups.

If you do not want to rename the images after the export, you can simply change the object ID of the given group before you start batch export!

To do this, right-click on the group of the icon, and select **Object Properties** from the menu. This will open the **Object Properties** menu, where the first variable is the ID. Rename your group there and hit the **Set** button at the bottom of the dialog window to save the ID for the object.

There are two rules worth mentioning about setting a new object ID:

- First, an ID is a name for the given object, not a filename. You do not have to name the group `icon1.png`, for example – just name it `icon1`. It will be still exported as `icon1.png` when using the **Batch Export** feature, as Inkscape will stick the format after the name. IDs have other functions in Inkscape too, exporting is just one of them.

- Second, every ID has to be unique in that document! If you have a group with an ID of `icon1`, Inkscape does not let you name any other group like that. To avoid any error, I usually name my icons by number, or by what it is depicting, for example, `icon-save`, `icon-user`, `icon-open`, and so on.

Saving SVG icons

If transparent PNGs are still the standard for mobile apps, then what about SVG files? Now, SVGs are the more elegant solution. They are, of course, scalable and editable for different devices, and pushed the bitmap icons into a fallback role. That means, the latter only gets used when the actual device cannot render SVG vector files properly.

If you are working in Inkscape and creating simple vector icons, it is only natural that you want to use these crisp vector images in your app or website design. Sadly, there is no native solution in the current version of Inkscape to batch export selected SVG objects as you can do with PNG files. But there is a comfortable way to manually save several icons in a few steps.

Follow these steps for each of the icons in the set to create individual SVG files:

1. Open the file containing the groups of the nine icons you created.

2. Create a new SVG document and open it in another window.

3. Select an icon in the source document and copy it into the blank .svg file.

    To be able to use the icon on a website, the icon has to fit the page and has to be positioned at the corner of the page. You can easily do this the other way around.

4. Select the icon, and push *Shift + Ctrl + R* to resize the page to the selection! (You could also achieve this from the **Document properties** dialog, but isn't this just so much faster?)

5. Go to **File** and save a copy from the current SVG file. Saving a copy instead of normal saving will help you to avoid accidental overrides.

6. Delete the content of this file, and start again from *Step 3*, copying a new icon into the empty file, resizing the page, saving a copy of it, and so on.

This method sounds more labor-intensive than simply batch exporting those PNG files earlier, but it is a simple process, and you can easily save plenty of SVG images in no time.

Exporting tip

There is a fast and automatic way capable of exporting multiple objects into several different file formats using Inkscape extensions. You can learn more about that in *Chapter 8, Pro Tips and Tricks for Inkscapers*!

Summary

In this chapter, you built matching icons for an icon set. At the start of the process, you learned about the theory of icon design, and how to implement that in a fast and creative way in vector. Then, you moved forward and created each of the line art icons you can use in the following chapters. This was a practice in not just designing icons with similar features but working with the **Path** tools and creating line art in Inkscape.

Adding and removing segments of a path while maintaining structure is a key part of icon design and vector design in general, and so is exporting the final design into different formats and exporting multiple files effortlessly. This chapter focused on icons, but as you can see, it also focused on creating images in quantity as efficiently as possible.

We will continue to explore more possibilities in the next chapter, where you will learn to use complex illustration techniques in Inkscape!

4
Create Detailed Illustrations with Inkscape

You might not want to reach your artistic goals as an illustrator right now. But you will still see that this chapter is full of resources and methods you can use to create complex work in Inkscape. Basic vector illustrations are a good way to learn Inkscape, such as the set of icons we just finished in *Chapter 3, Modular Icon Set Design with the Power of Vector*. But to provide reliable design services for a client, it is good to know the tools and rules to create detailed drawings as well. This is why this chapter focuses on creating a business illustration with many details and elements.

This project is not just about drawing. It is about using illustration to practice the use of shape and color and efficient work in Inkscape!

In this chapter, we're going to cover the following main topics:

- Sketching with Inkscape
- Adding colors and drawing your characters
- Drawing the laptop
- Finalizing your illustration

Sketching with Inkscape

As I stated in the previous chapters, whatever you're designing, sketching will help. It helps to clarify your ideas and set your focus on the goal of your creation, whether it be a logo, an icon, or an illustration. I always say that sketching your ideas on paper is a must, at least to lay down a few basic directions, to get the work started.

Sketching is important, but, in some cases, you might want to sketch in the program itself. Since Inkscape is a vector graphic program, you can easily draw a few basic shapes and mold them into their final form later in the process. Also, while creating illustrations, you are not as limited as, for example, with logo design.

When working on an illustration, you have more flexibility; you can use more details and add as many shapes and objects as you want, and you are free to experiment with colors and lighting. In general, you have fewer rules to follow, and you have a larger toolset to convey your message.

In this chapter, your task is to create an illustration for CloudUsers. The theme of the illustration is people using cloud services. We will draw the illustration together step by step while practicing with all the tools you'll need to create your own designs later!

This illustration will be more elaborate than the simple icons we created in earlier chapters. It will have more objects and more colors, and of course, it will tell a better story as well. The point of the illustration is to grab the attention of visitors to the website and get an emotional response.

In this case, what we want to show the visitor is that cloud services make communication and working together fun and easy, and that the people using CloudUsers's services are a friendly bunch. Actually, this is the message that a lot of IT companies want to pass on nowadays, and their illustrators have created a simple and clean *friendly corporate style* in response. For our illustration, we will use this clean and colorful style too!

Here is a very simple vector sketch I made earlier – you can follow my lead, or draw your own version based on the techniques you practice in this chapter.

Figure 4.1 – My quick vector sketch for the CloudUsers illustration

This very basic sketch is more of an inventory than an illustration. Its purpose is to help you decide what needs to be in the image. I decided to include a big laptop in the middle, some small people interacting and working, and of course clouds as the visual representations of cloud-based services. What colors should we use? I don't know yet! What is the final position of the elements? I don't know yet! What are the figures doing? I don't know yet! And that is all fine since this is just the first step on the way toward our final illustration.

Creating the first sketch in Inkscape

The goal here is really just throwing some elements down to put your first ideas on the screen. Therefore, you will only use basic shapes and simple colors. And don't forget to utilize one more thing: the cloud shape you created earlier, which is still there and ready to be used!

Let's create this first sketch together in Inkscape by taking the following steps!

1. Create a new document for your illustration, then open the file containing the basic cloud shape for the icons. Copy the cloud shape into the new document.

   Since we are not focusing on color at this stage, just assign any fill color to the cloud, and turn off the stroke color. I use some dull grays at this stage since they are easy to distinguish and don't interfere with the *artistic process*. This will not stop you from thinking about colors and that is great for speed!

2. Now draw a basic laptop. A laptop is just like a paper folded, two distorted rectangles sharing a common edge. Draw a gray rectangle and skew it to create the screen. Duplicate the rectangle and turn the object into a path. Then select the two top nodes of the path and move them to open the *laptop*. This will modify the second rectangle, sharing an exact edge with the first one while appearing to lie flat. See *Figure 4.2* for a visual representation of this.

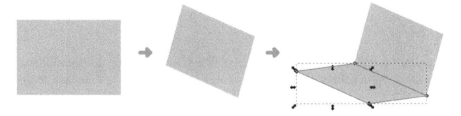

Figure 4.2 – Creating a basic laptop shape

> **Tip**
> Now, this is a trick I like to use when I create rectangles that share a common edge. It can be used for boxes, stairs, and so on. Duplicate the original rectangle and move the top nodes. This saves a lot of time since you don't have to align the two objects to each other!

3. Now add the clouds! Move and scale the cloud you had earlier to give a background to the laptop. Send the big cloud behind the laptop using *Page Down*. Pick a darker gray for it to remind you that it has to have a strong color contrast later.

4. Duplicate the cloud, scale it smaller, and move it a bit higher to the left. This will be positioned as a real cloud later and will provide a seat for one of our characters.

Figure 4.3 – Laptop in a cloud

The stage is set; it is time to add some characters. Since this is only the first draft, you will not work on any details, just add some simple shapes to represent the users. The goal is just to figure out the positions and the general scale of the objects in the illustration.

5.   Draw a small, rounded rectangle, and draw a circle above it, similar to the user shape we created for the logo.

6.   Duplicate the user twice. Move one of them to stand in front of the giant laptop. The second one will sit on the cloud in the sky – you can make the body shorter to signal sitting. And move the third character to sit on the edge of the big laptop. Refer back to *Figure 4.1* to see how I positioned them. Feel free to change the positions and play around with the size a bit in your version!

This is our flat vector sketch done. In the future, you will see that practicing this sketching technique will help you make faster decisions and make your design work with Inkscape much faster! Note that the sketch is not set in stone: you can and will modify it and change some parts later on. It is a flexible guideline for creating your final illustration.

Adding colors and drawing your characters

Now that our illustration is laid out with these flat shapes, it is time to go deeper and start adding those details. I know it may seem like it is easier said than done, but sometimes it is harder to imagine how to build up a vector illustration than actually doing it.

> **Tip**
>
> While building a basic sketch up to a complex vector illustration, follow the old rule of drawing: *instead of focusing on one area only, try to work on the image as a whole*. First, focus on the big shapes, then add some colors, then more and more layers of details, and finally, lighting and additional effects if needed. This way, the different parts of your illustration will be developed evenly. This makes it easier to keep things visually similar, and you will have an illustration where all the parts work together.

After sketching, the next step is to add colors while also adding more details to the illustration. Let's turn those simple placeholders into humans interacting in an interesting environment!

Adding colors to your sketch

We usually start coloring an illustration with a few colors that will support the mood and the message. We've all struggled with finding the perfect colors. In this case, you will create an illustration to match a previously decided logo and its color palette. So, our first colors are already decided: the blue (#00AAD4) and the turquoise (#2AC1B5) we used earlier have to be included.

The idea is to pick a few starter colors, and then we will mix these colors and recolor elements later on as needed. To create your own color template on the drawing board, just draw a small square, and apply the basic blue color (#00AAD4) as a fill color to it. Then duplicate this square and color it turquoise (#2AC1B5).

What other colors do you need for starters? There needs to be a lighter shade of the two basic colors, and there needs to be a lighter and darker skin color for our characters, as well as a darker variant of the original blue, to draw hair and add some contrast to our drawing. And finally, pick a type of vivid orange or red to brighten up the whole image a bit so it is not just shades of blue and turquoise.

Add a small square for each of these colors. Feel free to choose your own colors using the color wheel in the **Fill and Stroke** window. Keep an eye out for matching colors; move the small squares of your palette onto each other to see how colors relate. This is the vector equivalent of trying a brush of color on the side of your paper while painting. See the following figure for my colors and codes. (You don't need to copy the codes; I'm just showing the palette I created for your information.)

Figure 4.4 – The eight starter colors and their hexadecimal codes

To use this DIY palette, just select any object you would like to color and pick the color of the matching square using the **Color picker** tool! You don't need to save this as a custom palette, nor do you need to copy codes anymore.

You will not apply the colors to the sketch just now. Before coloring anything, let's design the characters in more detail.

Drawing the first character

Let's start drawing the first character! The user standing in front of the laptop is a good choice, as it is standing in a fairly simple pose and will be a good reference for the other characters as well.

Drawing a simple human character standing is an easy job. They have a head, a torso, two arms, and two straight legs. You can start designing your shape from the placeholder character shape you already created for your sketch:

1. Take the placeholder character and copy it to a new clean location away from your sketch. Then cut the bottom half by drawing a square over the figure using **Path | Difference**. Cut it a bit longer this time to add a slightly thinner body to the character. Then remove the circle as the head and draw a new rounded rectangle – a short pill shape – to be the head.

    Later on, when we add details to the head, we will keep it simple: no nose, no eyes, maybe only a simple mouth to smile and communicate some emotions. But for now, just draw the shape for the head, and position it above the body and slightly off-center. We positioned the head a bit off-center, so that the character is facing forward, toward where the laptop will be.

    Check out *Figure 4.5* for a visual representation of these steps. This is only my version and my favored cartoon character proportions. Feel free to experiment with the proportions of your own characters!

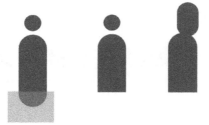

Figure 4.5 – Creating the body and the head using simple rounded rectangles

2. Next step: the legs. Use the **Bezier** tool and draw the legs with straight lines, starting from the hips. Just click with the mouse; do not hold the mouse button down!

    We just want simple straight lines this time. You can also create the feet now, as it only takes one extra click of the mouse to draw this simple triangle shape.

3. The same goes for the arm: it is a simple shape, so it will be fastest to use the **Bezier** tool and draw it with six easy clicks.

Refer to *Figure 4.6* for help with drawing the arm and legs. After you draw them, put them in their places next to the body.

Figure 4.6 – Drawing the arm and the legs with the Bezier tool

> **Tip**
> Try to limit the number of nodes you create, to keep your vector paths as simple as possible. The fewer clicks you make with the **Bezier** tool, the smoother your path will appear. The line will appear less shaky, and it will be much easier to reshape later.

The arm and legs are simple shapes, and you may simplify the hands too. You don't even need to draw any fingers. Why? Because although this will be a complex illustration compared to the icon you drew earlier, the characters and other elements in it will still be quite simple.

The combination of these simple elements will result in a complex image. So, you can create a hand with two simple shapes, and it will look like an expressive hand gesture when part of the bigger picture!

4. Draw a vertical, rounded rectangle – a pill shape – and cut half of it vertically with a square using **Path | Difference**. Then duplicate this half, scale it down about half the size, and flip it horizontally by pressing *H*. This will be the thumb.

Now move it so close to the other half that they are touching. Together they create a simple rounded hand shape, almost like a lobster claw. Rotate, scale, and position the *hand*, so it sits on the end of the arm of the character. Refer to *Figure 4.7* for visual help.

Figure 4.7 – Even simple shapes are enough to create a hand gesture on a smaller scale

The standing character is coming together nicely, and it is time to use some of those colors you selected earlier!

5.   Select the head and the hand shapes and use the **Color picker** tool on the *pink* square to assign the skin color to these shapes! Then select the torso and the arm and color them light blue in the same way. Finally, add the turquoise fill color to the legs, as shown in *Figure 4.8*.

Note that we are only using flat colors for now; you will add gradients and shadows later on too!

Figure 4.8 – Coloring the first character with the chosen colors

I have nothing against bald characters, but to add a bit more detail and personality to this one, let's design his hair. When drawing someone's haircut in vector form, you have to think about the hair as a mass. Picture it as a shape, not as a bunch of individual strings. In the case of this flat and geometrical character style, this is even more true.

6.   To create the hair, first, select the head shape, and duplicate it. Then select this duplicate and assign the darker blue fill color to it (#146478).

7.   Select the hair and rotate it about 30°, as seen in *Figure 4.9*. Now select the original head shape of the character again and duplicate that. You now have three objects covering each other – the original head shape, the darker and rotated hair shape, and the duplicated head shape.

8.   Select the top two shapes – the duplicate of the head and the darker hair shape – while holding *Shift*. Do not select the original head shape this time! Use **Path | Intersection** from the **Path** menu. This will create an intersection of the two selected objects. In this case, this will be a haircut that perfectly fits the shape of the top of the head. This is what I call the **sandwich method**.

9.   The hair of the character seems too flat now, so zoom in and add two rounded rectangles of the same dark blue color. Position them right at the edge of the forehead. This will create a hairstyle fitting a busy cartoon character!

Figure 4.9 – Creating the hair using the sandwich method

The sandwich method

When you need to cut a shape to the edge of another irregular shape seamlessly, use this method:

(1). You start with a base shape and a second shape that needs to be cut to the edge of the base shape.

(2). Position the second shape where you want to have it above the base. It has to cover part of the base.

(3). Duplicate the base shape, to have a third shape.

(4). Select these top two shapes and create their intersection with **Path | Intersection**, or *Ctrl + **.

Remember the order: original shape, above it the shape to be cut, then a duplicate of the original in the same position. Like a sandwich!

Why do you need to duplicate the original? Because using the **Intersection** operation deletes the two original shapes, and leaves only their intersection.

I use the sandwich method a lot during my work because it makes my illustration process much faster and more precise. Imagine freehand drawing a path above a shape that has to fit the edge of the original shape perfectly. It is possible, of course, but it takes much more effort and time than using this simple method. You can use it to create hair, fitting clothes, textures, lighting on objects, shadows, and so on. If you are illustrating, this is a very useful and versatile method! You will have plenty of opportunities to practice the sandwich method during this chapter.

The only thing we want to add at this stage is the arm on the other side of the character. Of course, only the character's elbow will be visible; still, this will emphasize his gestures, and he will appear less flat.

10. Use the **Bezier** tool and draw a simple triangle for the arm. Select the **Path editor** tool, then press *Ctrl* and left-click on the node at the elbow. This will turn that corner into a curve, and the arm will look more believable. Now select the arm and send it behind the body of the figure, pressing *End* once or *Page Down* a few times.

The rear arm has to look like it is behind the body, so it needs to be visually separated from the torso.

11. Select the rear arm and change its fill color to the original blue you created earlier (#00AAD4).

12. There are still a few corners around our character that we should soften up. Use the same method that you used on the rear elbow, but this time use *Ctrl* and left-click with the **Path editor** tool on the nodes in both knees and the elbow of his front arm.

This will create a more organic look for the character. Keep the feet and the hand as simple shapes for now; it will support the simple, friendly look.

Figure 4.10 – Notice how the slightly curved joints create a more organic look

The first character is ready for now. It will be great to set the style and proportions for the remaining characters against this character. If this seems too flat and simple for you, don't worry; we will revisit this character later in this chapter when we add lights and shadows to the elements of our illustration.

But before moving forward, select all the parts of the character and group them by right-clicking and selecting **Group**, or select **Object | Group** in the top menu. But of course, the fastest way, as always, is using *Ctrl + G*. Why is this necessary? You will learn about this in the following section.

Using groups and layers to keep your file organized

Before drawing the next character, let's pause a bit and learn about organizing your work in Inkscape. The most effective tools for that are, without a doubt, groups and layers.

We already used groups to hold together the icons we created in *Chapter 3, Modular Icon Set Design with the Power of Vector*. Grouping is still useful even if an average icon only consists of 5-6 individual elements. This number in a complex illustration can be anywhere from 50 to hundreds or even thousands! And believe me, when objects start to pile up in a document, you need a method to create some order before it all falls into chaos.

The list of layers is a well-known sight to anyone who has ever used design software. In Inkscape, it is not so obvious, though; you have to press *Shift + Ctrl + L* to show the **Layers and Objects** panel. After that, layers work the same as in any other software: whatever you draw or copy on one layer will be held on that layer. Layers can be named, and you can stack them over and under each other too.

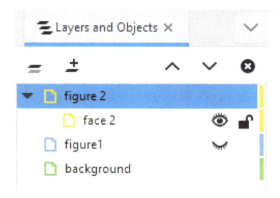

Figure 4.11 – The layer tab in Inkscape 1.2

The question is should you use layers or groups while working in Inkscape? Well, in answer, both. Here is the personal opinion of the writer of this book: layers are actually groups in a list, and I usually work better without them. Groups are faster to create and easier to manipulate on the fly. Hence, I will mostly use groups in this book.

But there are some advantages of the layer system that I do not want to skip and do appreciate from time to time! First of all, you can apply different blend modes to layers, creating a visual effect you could only do with a lot of work otherwise. You can set a layer's **Blend** mode to multiply, darken, lighten, and so on, affecting all the layers underneath. This is a great tool!

Second, layers can be locked and hidden (note the lock and eye icons in *Figure 4.11*). When a layer is hidden, Inkscape does not display it, thus sparing memory and keeping your computer working fast.

Groups and layers are both great ways to organize your work. It does not matter which one you prefer, just please get used to organizing your elements! People who work with your vector files in the future will be very glad about this – including future you!

Let's move forward and design the remaining characters for our illustration. Don't forget to practice your organizing skills!

Drawing the remaining characters

Designing the second and third characters will be an easier task now that the first character is finished. Remember, we are drawing in vector form now, so we will salvage and reuse any useful elements previously drawn in the current project! As a designer, our intention always has to be to create original content and to avoid copying someone else's or our past work.

But one of the appeals of working with vector graphics is how easily you can reuse and redesign existing shapes and parts. And we will practice exactly that in the following sections.

Drawing the second character

In my sketch, the second and third characters have similar poses. One of them is sitting on the edge of the big laptop, and the other is working up in a *cloud*. Let's continue our design with the latter.

Before even drawing the first object for this character, let's see what we can salvage from our existing design! Since the character will sit on the cloud, we do not have to draw that for sure:

1. Create a copy of the cloud you used earlier in the sketch. The design of the first character is a good starting point, too. Copy the body and the head of the standing man here too. We will not reuse the legs and the arms of that character, but their shape can work as design guidelines for this new character.

    The cloud from the sketch is gray, and the torso of the man is light blue. Let's fix those colors before going forward.

2. Use the **Color picker** tool on your color palette, or on the torso of the man, and set the cloud's fill color to light blue. Now, to create contrast and some variety in your illustration, select the torso of the new character and apply the blue color to it from your premade color palette (#00AAD4).

Figure 4.12 – Using and recoloring elements you created earlier

A simple color change is not enough of a difference to create something new. Besides, if we want to create a female character, then we should change the proportions a bit too.

3. Use the **Path editor** tool to select the top three nodes of the torso and scale down this part of the path to make the shoulders narrower. Also, move the nodes lower so the torso is shorter than the one of the standing man. After that, move these nodes to the right to skew the woman's upper body a bit and make her appear like she is leaning forward.

4. Select the head and move it where the shoulders are. While selected, press *Home* or *Page Up* to bring the head forward. This character will look toward the viewer, not away like the previous one. So, the head shouldn't hide behind the shoulders. Rotate the head a few degrees so it appears to be looking down at the laptop. It might seem like her posture is off now, but it will all come together soon.

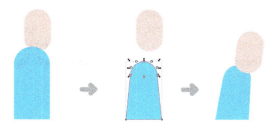

Figure 4.13 – Changing the position of the top nodes and the head to create a different posture

5. Again, before going into details, we will define the whole shape of the character. Time to add her legs. You can follow the style of the legs of the other character, simple shapes for the legs and feet. But this time, the character is sitting. Use the **Bezier** tool to draw one leg of the character with straight lines. Use the dark blue (#146478) as the **Fill** color.

6. If you are satisfied with the shape and position of the leg, pick the **Path editor** tool, then press *Ctrl* and click on two nodes to curve them. One of them is the node at the corner of her bottom, and the other is on the opposite side, the corner of the top of her knee. These two curvy nodes are enough to make the character look softer, and a bit more relaxed.

    Why do we curve the nodes of the leg now and not later? Because this way, you only have to draw one leg and still end up with the image looking more natural. You will now duplicate the first leg and modify it with the **Path editor** tool.

7. Select the nodes building the lower leg and move them to the right. This will elongate the leg a bit, but it will help create the pose. Keep these nodes selected and skew them so her leg is less bent. Move the nodes a bit higher to create a bit of distance between the two feet. She is now sitting in a comfortable chair with her legs crossed, working with a laptop on her thighs. See *Figure 4.14* for my version.

Figure 4.14 – Draw a leg with straight lines, then curve it and duplicate it into position

The design of her laptop can be really simple. Just two shapes, almost like a laptop symbol. No elaborate keyboard or screen content is needed since it is facing toward the character.

8. Use the **Rectangle** tool and draw a light gray horizontal rectangle for the body of the laptop. Then draw another rectangle above the first, sitting on the top edge of it. Select and skew this top rectangle a bit so the laptop seems open, with the screen tilted backward at an angle. Select the shapes and group them by pressing *Ctrl + G*. Move the simple laptop into position.

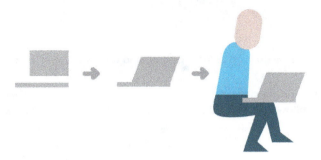

Figure 4.15 – Create the simplest laptop ever made

9. The next step is to draw her arm reaching toward the keyboard. Simply use the **Bezier** tool and draw an arm shape with straight lines. The top of the arm is cut off straight now, but that is fine: it creates the illusion that she is wearing a t-shirt! You do not have to draw the t-shirt sleeve; the viewer will still understand it is there. And because it is a t-shirt, the skin on her arm must be bare. Use the **Color picker** tool and pick the skin color from the character's head.

It is also believable that her hand is covered by the screen of her computer while she is working on the keyboard. Lucky for us, as it means we do not have to draw her a hand this time.

Figure 4.16 – Add an arm with no hand behind the screen

10. At this level of detail, the last things to add to the character are the hair and a smile. We will aim to draw thick, wavy hair held together in a bun – in a simple geometric form. To create the hair, we will use the **Path Boolean** operations again. Follow the exact steps as shown in *Figure 4.17*.

11. Draw a circle slightly smaller than the head. Send it behind the head with the *End* or *Page Down* key and move it upward and to the left. This will create the base of the hair. Color the circle the vivid red we chose earlier (#FF5555).

12. Select the red circle and duplicate it. This time, you will use another version of the sandwich method you learned earlier. We will use the duplicated circle to cut off a part of the head of the character. The head with this missing part in front of the original circle will create the illusion. It will seem that the hair is in front of the forehead, covering a part of it. To create the effect, move the duplicated circle more to the left and upward. Then, select the head shape and the duplicated circle, and go to **Path | Difference**, or use the *Ctrl + -* hotkey.

13. If you are not sure how much to cut off, use the **Outline Overlay** display mode to see the parts that are covered.

Tip

Use the **Outline Overlay** display mode to see all the elements with their outlines and their fill colors visible! It is a new feature from Inkscape 1.0 that helps a lot when shapes are covered and when you are working with a lot of objects and are having a hard time distinguishing between them. Activate this mode from the top menu by selecting **View | Display mode | Outline overlay** or use *Ctrl + 5* to quickly cycle through all the display modes. After finding and selecting the object you are looking for, you can simply go back to normal view the same way.

---

14. Draw a small red rounded rectangle and rotate it 45° to create the bun. Move it close to the hair and scale it to its final size.

15. After you are done with the hair, add a smile to the character's face. Simply draw a white circle and turn it into a semi-circle using the handlers with the **Circle** tool. If you just position this white semi-circle on the face, you will notice that the expression is somewhat off (as illustrated in the last steps of *Figure 4.17*). Since the character's head is tilting down, the smile should be tilted as well. Select and rotate it. And since she is facing the laptop and not the viewer, we should also move the smile closer to the right side of the head.

16. Select all the parts of the head and group them together by pressing *Ctrl + G*. Then move the head onto the body and group it with the other parts of the character.

Figure 4.17 – Drawing hair in a bun and adding a smile to the character

17. This character is in a sitting position, so she needs a chair to sit on. Move the group of the character onto the cloud and put it into its final position. We will create more of a seat for her later during the illustration process.

Figure 4.18 – The second character on her cloud

Congrats, the second character is almost finished! We will add more details later on, as we will add lighting and some interesting elements to the whole illustration. Let's move on to the third character, then build the large laptop and the environment!

Building the third character

Creating the third character will be the easiest of them all for two reasons. First of all, you are *warmed up* by drawing the characters before, and second, almost everything you created earlier can be used in this character. This character will be a bald male, sitting with his phone on the edge of the big laptop.

Let's start the drawing process with the same step as before and look around the drawing board for elements we can recycle!

1. Since the character is a male, you can select the upper body and head of the first character, the one that is standing. We can also use his hand and copy that as well. Then select one of the legs of the second character, who is also sitting, and copy that here too. There are no rules to this; just look for elements that are easier to copy than create from scratch.

Figure 4.19 – The elements that you can reuse from the previous characters

Of course, we cannot just put the elements next to each other and call it a day. You have to modify and shape them into a new character. It is already visible at first glance that the upper body is too long, and seems too big for the legs.

2. Using the **Path editor** tool, select the top nodes of the torso, and move them a bit lower. Also, scale up this part of the path of the torso. This is the opposite of how you created a leaner feminine character before. Now you are aiming to create a man with a bigger frame and different posture. Playing around a bit with the character shapes will create more life and diversity in your illustration.

3. To create a different sitting position than before, select the nodes in the feet and pull them to the left until the lower leg is close to vertical but bending under the character. The thigh is also thin; editing its nodes with the **Path editor** tool is a good idea. Just select the nodes at the bottom and at the knee pit and move them lower. See my version of this step and the previous one in *Figure 4.20*!

Figure 4.20 – Modifying the nodes of the torso and the legs for a new character

4. Duplicate the leg to create the other leg. This time select the nodes of the thigh and the bottom of the character and simply delete them. This will cut off the rest of the leg and keep the lower leg and the feet. Position this leg onto the other one so the top of the leg is not visible. Make the leg longer with the **Path editor** tool and skew it a bit to create a relaxed sitting position for the character.

> **Tip**
>
> When using the **Path editor** tool and deleting nodes of a path, things can go strange in a second. The shape can grow weird bulges and seem out of control, as seen in *Figure 4.21*. If this happens, relax, as this is a common incident. Inkscape just keeps calculating the direction and distance of the path, as it should. Still using the **Path editor** tool, look for the node handles that are causing the trouble. Grab the handles and pull them closer to their nodes. This will refine the shape. It seems alarming at first, but the more you use the **Path editor** tool, the better you will get with it!

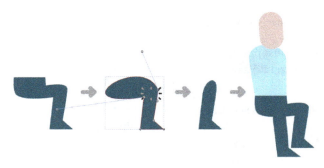

Figure 4.21 – Deleting the nodes of the leg to build the other leg

Before adding the arm, let's apply different colors to this character. After all, you do not want to use the same colors on all of them (unless they are all wearing the same uniform). The aim is to play with the variations of the colors but not overdo it.

5.   To create more character variety, use the **Color picker** tool and pick the darker skin tone (#AE8C86) for the head. The trousers and the feet can stay the same dark blue color, but for the torso, we will use the original turquoise color (#2AC1B5). This will build a smooth contrast with the character's skin color and later with the background too.

Now to create the arms. This character is sitting and talking on his phone while waving his hands in excitement. A lot of us use hand gestures in phone calls, even though they are useless in this situation. This is normal human behavior and adds to the illustration. So, one of his arms has to be bent and holding the phone. In this pose, the easiest way to draw it is to only draw the forearm, and not the upper arm.

6.   Draw a long rounded rectangle, and color it to his skin tone with the **Color picker** tool. Then, scale and position this rounded rectangle between the face and the *elbow* position of the character, as shown in *Figure 4.22*.

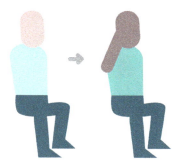

Figure 4.22 – Recoloring the character and adding an arm

7. Drawing a simplified smartphone in Inkscape is easy: pick the **Rectangle** tool and draw a small rectangle with slightly rounded edges. Color it the darkest blue color we use in this illustration, the same as the trousers of this character (#146478). Now put the phone *into the hand* of the character: scale it and rotate it into position, then press *Page Down* to order it under the rounded rectangle of the arm.

8. Although we are not using the whole hand here, copy the *thumb* created earlier and position it under the phone, as shown in *Figure 4.23*. Color it with the skin color as well, and it will give the perfect illusion of someone casually holding a phone. As the color of the head, the arm, and the thumb are the same, you do not have to be very precise when fitting the phone and the hand; a small overlap is fine.

Figure 4.23 – Building the phone and the hand

9. Draw the other arm. Use the **Bezier** tool, and create the simple arm shape we created earlier, just in another position. Press *Ctrl* and click on the node at the elbow, to turn it into a curve.

10. We will create the t-shirt effect by dividing the arm into two parts: the sleeve and the bare forearm. To do this, first, use the **Bezier** tool, and draw a shape right over the arm where you want it to be divided. Then, select both this shape and the arm, and click on **Path Division** in the top menu bar. The shape will disappear and will cut the arm path into two. Use the **Color picker** tool, and color the part on the left to the color of the torso, while the part on the right has to be skin-colored. See *Figure 4.24* for more help!

Did you copy the hand of the other character here earlier? Great, now position it at the end of the gesturing arm of this character, and color it to match the skin color of his arm.

Figure 4.24 – Draw the arm, create a t-shirt sleeve, and position the hand

The only thing we're missing is a great smile to show that the character is in conversation! You could copy the one from the woman you drew earlier, but it is such a simple shape that it may be easier to draw it here.

11. Just draw a semi-circle with the **Circle** tool and color it white. Then position and scale it onto the character's face. Take care to position the mouth under the hand and phone using the *Page Down* key.

12. The third character is complete for now. Select all of the parts and create a group of them with *Ctrl + G*.

Figure 4.25 – The third character finished with a smile!

In this section, you drew three characters using the best things Inkscape can offer! You used simple shapes to build expressive characters. You created an easy color palette for your illustration and applied it to your characters. You learned about groups and layers and about how they can help your work. You learned my sandwich method to create perfectly fitted parts for your illustrations. And you managed to finish all three characters effectively, reusing and re-purposing their elements in a modular way.

And now that all the characters are set up, it is time to focus on the background of the illustration.

Drawing the laptop

So far, you have created a standing character and two others seated based on your initial sketch. Now, we will create the environment for them to give meaning to the whole illustration. We will start with the giant laptop as the central object in the story, then add more background elements. You will start with the big shapes and slowly work your way down to the smaller parts.

You created very simple *laptops* earlier: one for the woman working in the clouds and one for your sketch. The former will remain simple, as it is only a small object; the focus is on the character using it. The latter, on the other hand, is a placeholder for something far more important! The huge laptop in the middle connects the characters and offers a great backdrop for the illustration. It is evident that this laptop needs a bit more attention than the other one.

Also, this giant laptop is open in a different position than the small, simple one. Both the keyboard and the screen are visible. This means that you have to create content for the screen and draw a keyboard! This might seem like a very complex task, but you can make it painless with some planning!

Building the keyboard of the laptop

You could do your own calculations and draw each of the elements of the laptop in its final position manually. But why would you? The biggest benefit of using Inkscape is the flexibility of working with vector elements. Coloring, scaling, or distorting objects or a group of objects takes only a few clicks. This is exactly what you will practice now!

The plan is to create the rows of keys and the base of the laptop first in a flat, top view. Then group the elements and skew them into position, matching the view of the elements to your sketch.

As this is not a realistic illustration, we will draw an abstract keyboard. There will be no letters and numbers on the keys, and the keys will be bigger and fewer than in reality. A simplified keyboard, but complex enough to create the illusion:

1. It all starts with the first key! Draw a square with slightly rounded corners and color it blue (#00AAD4). Now create a duplicate of it using *Ctrl + D*. Select and move this duplicate next to the original square. Now select both of them and create a new duplicate. Move these two new keys next to them too! Now select all four keys and duplicate them. When you have eight keys, select them, and hit duplicate four more times. This will create 40 keys.

2. Your keys might seem scattered around, but don't worry. Go to the bottom of the **Object** menu in the top menu and click on **Arrange**. This will open the **Arrange** dialog window. Select all 40 keys and arrange them in 4 rows and 10 columns.

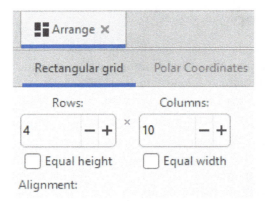

Figure 4.26 – Arranging the squares to create a 4x10 grid

3. Set the spacing higher than **0**, to see a gap between your keys. Hit the **Arrange** button, and if the gap is small enough, raise the **X** and **Y** distance and arrange again until you are satisfied.

4. A regular 4x10 grid of squares looks very ordered, but it is not a keyboard yet. To have that familiar look, you need to add some special keys. For this, delete three keys in the bottom row, as shown in *Figure 4.27*. Then select the key next to the gap you just created, and using the **Rectangle** tool, elongate it horizontally to create the spacebar. You can also use the **Selection** tool to do this, since the key is an object, and will keep its curved edges upon transformation.

5. Then delete the second and the last keys from the third row. One will give space to the *Shift* key, and the other one to the *Enter* key. Elongate the first key horizontally as before. Finally, elongate the last key on the second row vertically to fill the gap with a vertical key.

6. Select all of the keys, and group them with *Ctrl + G*. Your illustrated keyboard is completed. It is not perfect, and has fewer keys than a real keyboard, but the viewers will recognize it.

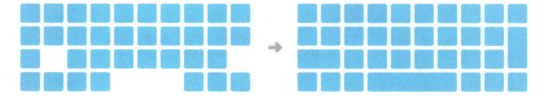

Figure 4.27 – From simple grid to keyboard

7. To add a touchpad to your laptop, you need a bigger rectangle under the keyboard. Just select the spacebar key from the keyboard by pressing *Ctrl* and clicking it, then copy it outside the keyboard group. The width is correct for a touchpad but you need to scale it about two times higher. You can keep the same color just for now.

8.   The only thing you need to add now is one bigger rectangle under the keyboard and the touchpad. This will be the base of the laptop. Draw it bigger and wider than the whole keyboard. Then select it and send it to the back with the *End* or *Page Down* keys. You can color it gray or light blue, for now, just to distinguish it from the other elements above it. To position it to the keyboard group, select the group and the base rectangle, and align them to their centerlines using the **Align** dialog window.

Figure 4.28 – The keyboard and the touchpad on the base

9.   Select the keyboard, the base, and the touchpad, and create a new group out of them. The lower part of the keyboard is ready.

It's now time to create the screen.

Drawing the screen

Drawing the screen of the laptop is fairly easy if the screen is turned off. But we want the screen turned on and showing something interesting. The easiest thing to show on the screen is yet another character:

1.   First, duplicate the same base rectangle you added under the keyboard. Just double-click the group to enter it, then copy the base rectangle and paste it outside of the group. This will be the frame of the screen, the top part of the laptop.

2.   Duplicate the frame, and scale it a bit smaller while holding *Shift*. Holding *Ctrl* would scale it proportionally, but you don't need that now. Holding *Shift* will scale the object around the middle point. This saves you time since you do not have to re-align the screen to the center of the frame. Color the screen brighter than the frame, for now, just to see the difference.

3.   Add a small dark circle to the frame above the screen. This will be like a small camera lens, but it will also add some more detail to the laptop.

Figure 4.29 – A simple screen, a frame, and a camera

4. Now, to make the screen alive! We won't leave it blank but rather add a fourth character talking on the big screen. You will do this in a blink, reusing the existing characters. You know the drill already: copy the torso and head of the first character you created.

5. This torso is in a neutral shape and position but too tall for the screen. Draw a rectangle to cover it partly, finishing just under the shoulders, and cut the body off using **Path | Difference**.

6. Add a white semi-circle to create a smile on the character's face. Now, to add a different haircut, draw a rounded rectangle and set it behind the head. This will look like straight, long hair. Color it in dark blue, as we used earlier. We will use this color and not black or dark brown because we want to keep the color palette of the image coherent. This dark blue is the darkest color in the illustration.

7. Add a smaller ellipse, the width of the head, to create a fringe for the character. Position it over the head. Select the fringe, the head, and the smile, and rotate them a bit simultaneously, to tilt the head.

8. Now rotate the hair slightly in the other direction to add a bit of head movement to the image. Change the shirt color to red (#FF5555), then select all the parts and group the character.

Figure 4.30 – Designing a new character for the screen

9. Position and scale the character over the screen, fitting the bottom of the character to the bottom edge of the screen. Select the character and the screen part of the laptop and create a group of them with *Ctrl + G*.

Both the top and the bottom parts of your laptop are created now and you have the keyboard and the screen ready to assemble.

Assembling the laptop

The next step is to connect the two parts and make them look a bit more like an actual computer:

1. To start, create duplicates of three things: the top and bottom parts of the laptop and the laptop vector sketch you created at the start of this chapter. Create copies of the keyboard and monitor groups to be able to go back to the original flat versions and change something if you need it. Trust me, redrawing an element in a rotated and skewed group is no fun.

    We will use the **Selection** tool to skew and rotate the bottom part of the laptop into place. How do we do that fast? The following method will help to skew this group of elements without guessing directions and creating unwanted distortions in your drawing.

2. Locate the topmost and lowest corners of the laptop sketch base and use vertical guidelines to mark their positions (I also marked them with small arrows in *Figure 4.31*). Then position and rotate the lower part of the laptop to match the corresponding corner to the topmost corner of the rectangle in the sketch.

3. Then, still using the **Selection** tool, scale the group vertically to match the bottom corner of the sketch too. Don't worry about keeping the proportion; for now, the point is to skew this group according to the viewpoint of this illustration.

4. Two corners are matching; now, scale the object horizontally to fit the two remaining corners as well. Then slightly skew it vertically a bit more if needed. We do not want an absolute match with the sketch; we just want to create an illusion of perspective here. This is not a real perspective graphic; this is also the reason why this method works easily here. Delete the bottom part of the sketch laptop as you do not need it anymore.

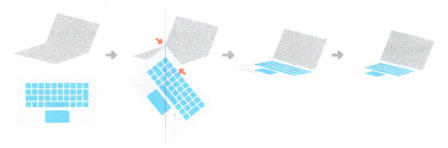

Figure 4.31 – Skewing the laptop keyboard cleanly into position

To elevate the laptop from a 2D object into a more realistic shape, it needs some thickness.

5. Double-click the group, select the base rectangle of the laptop, and duplicate it. Pick a darker gray color for this rectangle. Pressing *Page Down*, send the duplicate behind the original. While still selected, move it slightly lower to create some thickness for this part of the laptop.

6. Notice how there are gaps between the corners on the sides? Draw one small rectangle on both sides and align them to the corners of the laptop. Then select these small rectangles and the base and merge them together with **Path | Union**.

Figure 4.32 – Duplicating the base and filling the gap on the corners to make the laptop thicker

The keyboard and the lower part of the laptop are finished. Of course, now you have to go through the same routine for the upper part. Except it is not exactly the same, but slightly different this time.

7. First, take the screen and position its bottom-left corner to the topmost corner of the lower part.

8. Then, use the **Selection** tool and, holding *Shift*, pull down the arrow on the right side of the screen to skew it. Our aim is to match the meeting edges of the lower part and the upper part of the laptop.

9. If the edges are parallel to each other, grab the skewing arrow on the top, and pull it right while holding *Shift*. This will make the screen look bent backward a bit while keeping the edges still parallel to the keyboard.

10. Finally, if the screen is too big or too small, rescale it by holding the *Ctrl* key. This will keep every edge parallel to how you set it up before, but you might have to reposition the group to sit on the keyboard perfectly. Take a look at *Figure 4.33*. I signaled where to skew the group and in what direction.

Figure 4.33 – Putting the laptop screen in place by keeping the edges parallel

11. The top part is always thinner than the bottom, but it still has some thickness. So, double-click the group, and select the frame of the screen. Duplicate it, and color it in the same dark grey you used for the thickness of the lower part.

12. Then move it slightly to the right and down, just to create a small distance from the edge. Send it to the back using the *Page Down* key.

13. If there is a gap at the rightmost corner, turn this rectangle into a path using *Shift + Ctrl + C* or select **Path | Object to path** and slightly lift the nodes with **Path editor** to close the gap. I use this small trick a lot when the corner is rounded, and it does not have to be absolutely precise straight lines since it is just to close one gap.

Figure 4.34 – Duplicating the frame to make the screen thicker

There, the main laptop seems completed now. It is bigger than it looks; it will be the main background component and the stage to host our characters.

14. Move the laptop and the three characters into a new empty space and set the characters using your original vector sketch as a guide. (See *Figure 4.1* and *Figure 4.35* for reference.) Scale the laptop up while keeping the character sizes similar to each other. Don't forget to hold the *Ctrl* key while scaling to keep proportions.

Again, this is why I prefer to use groups rather than layers in my work: just select the character you want to move and easily move and scale it into its final position. The woman sitting on the cloud can be put in the top left, the standing character can go in front of the laptop, and the third guy can sit on the laptop while talking on his phone.

Figure 4.35 – The characters and the laptop in their final position

In this section, you created a laptop in a simple and logical way by creating a flat design and skewing it into its final position. It is a straightforward process, from creating the rows of a keyboard to adding another character to the scene by displaying them on the laptop screen.

The characters are now set in the scene, but there are still a few things to fix or add. Let's move on to the last phase, and add more background elements, shadows, and lighting before the finishing touches.

Finalizing your illustration

Of course, you could stop at this point, and claim the illustration is finished. Or you could continue to think about it and add more details, more lights, different colors, and so on.

From basic shapes to detailed illustrations, what is the difference? The difference is not being afraid to use your knowledge and patiently drawing shapes over shapes! We will upgrade our illustration from basic to complex with shadows and lights and details layered on each other.

Every illustration is built on simple, solid foundations, and then we use more elements to better express the message. And this is what makes an illustration look more professional.

Adding more background elements

The elements we are about to add to the illustration have a dual role. Their first role is to add more to the story. I know this is a long chapter about characters and laptops, but our main goal is still creating an illustration for the CloudUsers website. So, we need to add elements that convey the intended message and show the viewer what CloudUsers is about. There have to be elements referring to cloud services, communication, and connection.

The second role of the elements is composition. We need these visual elements to connect the main subjects of the illustration and *hold everything together*. If you look at what we have got so far, you might notice that the parts of our illustration are a bit separated. They need a background and small elements to fill the space when needed:

1. The first thing to add is the big cloud that was already present in your original sketch. Copy that cloud here and send it behind the laptop.

2. Scale the cloud up, also like in the sketch. Position the bottom line of the cloud to the corner of the laptop and then set its fill color to blue (#00AAD4).

Figure 4.36 – The big blue cloud behind the laptop

The image could also use some up and down arrows as signals for uploading and downloading data to the cloud.

If you created the icons in *Chapter 3, Modular Icon Set Design with the Power of Vector*, look for the one featuring a pair of up and down arrows. If you did not draw the icons then just go back to *Chapter 3, Modular Icon Set Design with the Power of Vector*, find the *Creating the icon for Cloud storage* section, and create just one of the simple arrows we drew there.

3. Pick the arrow and copy it into your illustration here. Duplicate the arrow with *Ctrl + D*, then flip one of them vertically by hitting *V*.

4. The arrows probably look very small compared to your characters and laptop, so scale them up! Make them about as tall as the standing character. Position the down arrow right on the cloud where the character is sitting and working. And place the up arrow on the right side, pointing up from the big cloud.

5. Color the arrows using the turquoise from the logo (#2AC1B5). Look at *Figure 4.37* to guide you in scaling and positioning:

Figure 4.37 – Add the up and down arrows to the clouds!

To add more movement to the image, and a feeling of communication, let's create some speech bubbles! They are cliché, but they will work well here!

6. First, create a horizontal rectangle and slightly round the corners of it. This will be the base of the speech bubble.

7. To add the *fin* to the bubble, you need to draw a small right-angled triangle. The easiest way to do this is using the **Bezier** tool. Hold down *Ctrl*, draw a downward vertical line, then an upward diagonal line at a 60° angle, and then close the triangle. It is OK if the top of the triangle is not perfectly horizontal.

8. Place the triangle under the base of the bubble, select both of them, and create a group of them. Color the bubble light blue by using the **Color picker** tool on the shirt of the standing character (#9EECFF).

Figure 4.38 – Creating a simple speech bubble

9. Now duplicate the speech bubble twice so you have three in total. Add one to the lady talking on the screen. Take care that the *fin* of the bubble is pointing toward her! If you need to flip the bubble, just hit the *H* key for a horizontal flip.

10. Place another bubble above the man talking on the phone. Make this bubble smaller and flip it horizontally. With this, it will not cover up the character on the big screen while still pointing in the correct direction.

11. The last bubble should go above the laptop of the woman working on the cloud. Make this bubble smaller than the others to create some variation. Check out *Figure 4.39* to see the placement of the bubbles!

    The last objects to add are some floating circles. What are they for? They could symbolize *data packages* or pieces of clouds, but their main reason is to fill the space a bit more and help create a soft, friendly environment.

12. Select the **Ellipse** tool to draw 4-5 small perfect circles on the scene, two of them on the right, near the floor, to fill in that blue space of the big cloud, and two or three of them in the air above the big laptop. Make some circles smaller than the others to make them look more playful.

13. Color all the circles light blue for now. Again, you can use the **Color picker** tool and pick the blue of the shirt of the standing character (#9EECFF). Look at *Figure 4.39* to see the placement of the circles!

Figure 4.39 – Placing speech bubbles and adding some circles

All background elements are added, and the illustration is visually balanced and gets the message through. In the next step, you will add a final touch to each element to make the illustration really pop!

Adding basic shadows and lighting

Adding even basic shadows and lights to your elements simply elevates the whole illustration. They define space and relationships between the different image elements; they can guide the viewer and even add new layers of meaning.

In this case, the shadows and lights will make these characters look like they are in the same space. We will add new objects to each character to create additional shadows and lights on them. The fastest way to achieve this is using the *sandwich method* that you learned about earlier in this chapter while drawing the hair of the first character.

Let's jump into it, adding some shadows to the first character:

1. With the **Selection** tool, double-click the group of the standing character to get into the group. Now we apply the sandwich method: select the torso and duplicate it. Select this duplicate and move it to the right and lower it a bit, still overlapping with the original torso.

2. Then select the torso and duplicate it again. Select the two duplicates and choose **Path |
    Intersection** from the top menu or hit *Ctrl + *.

    This will create an intersection of the two objects, which will be a great element to add light to the character. How? By darkening the original torso object under it.

3. Select the original torso and color it the original blue you created in the beginning (#00AAD4). It will be darker, and the light part is clearly visible now, but the contrast is huge. The difference should be milder. Let's ease this color difference a bit with **local color mixing**.

> **Tip: local color mixing**
>
> This is the time to do some color mixing. The eight colors you previously selected are a good start, but you often will need more. To keep the new color in the range of the original palette, use local color mixing. To use this method, select any object, and using the **Color picker** tool, hold the mouse button and pull away from the original picking point. This will draw a circle around that starting point as long as you hold the mouse button. Any color values in the circle will be applied to your object, ending up with an average of color and opacity values. I use local color mixing in illustrations whenever I want to get a new color mixed from the color values of the surrounding elements.

4. To create a milder blue, select the original torso. Then, with the **Color picker** tool, draw a circle over a small portion where the original torso and the light part meet.

5. Repeat and refine this process as much as you need! Try to find a color that is lighter than the blue you started with, but dark enough to create a nice visible contrast with the light blue. See *Figure 4.40* for the steps and my color choice! The background is hidden for clarity.

Figure 4.40 – Adding the shadow to the first character's torso using the sandwich method

You just created the first shadow of your characters! Now, you will repeat the exact same method for all the torsos and legs for all three characters. First, use the sandwich method to cut the pieces into the places they need to be, then use color mixing to have a coherent color palette for shadows and lights.

6. Let's move on to the legs of the standing character. If the light is coming from the same direction, the light part has to be on the same side too. You will just add a simple light stripe to his right leg now. This time use the **Bezier** tool and draw a shape that follows the curve of his leg slightly and exists at the top and the bottom of the leg shape. It does not matter how the shape looks outside of the leg shape since it will be cut anyway. If confused, check *Figure 4.41* for reference.

7. After the shape is drawn, select the original leg shape and duplicate it. Your sandwich is ready. Use **Path | Intersection** and cut off the parts that are not needed.

8. Select the original leg shape and set its fill color to a darker turquoise. You could simply mix a new color, or to work faster, set it to the dark blue of the hair using the **Color picker** tool, then mix your new turquoise shadow color on the spot.

Figure 4.41 – Adding light on the leg, while making the original leg shape darker

The character appears to be almost 3D now; the shadows on his torso and legs add extra detail. This is exactly the visual effect we are looking for.

Let's recreate the same effect for the other two characters as well!

9. Focus now on the character sitting on the edge of the big laptop. Double-click on the group and select his torso. Duplicate the torso with *Ctrl + D*. Since this will be a shadow on the turquoise color, you can use the **Color picker** tool and assign the same dark turquoise shade you mixed earlier for the trousers of the standing character. You can spare some time not thinking about finding a matching color and just use what is already available.

10. Move the torso duplicate to the left and lower it a bit. Duplicate the original torso again and use the sandwich method. Select the two duplicates and choose **Path | Intersection** or hit *Ctrl + **.

11. For his legs, use the **Bezier** tool, and draw a shape for each of his legs. In *Figure 4.42*, I started with the forward leg, drawing a light blue patch covering half of it. Then I duplicated the original leg, selected the duplicate and the path I just drew, and then created the intersection.

12. Then repeat the same for the other leg. Draw a shape over it to create the lighter part, then duplicate the original leg shape and create an intersection of the two objects. Since his trousers are dark blue, use local color mixing and create a lighter blue that is closer to the original shade.

Figure 4.42 – Adding shadows first to the torso, then to the legs of the sitting character

Now you are done with the sitting guy too! If you zoom out and look at the picture as a whole now, you might notice that the blue keyboard and the turquoise shirt of this character are not creating enough contrast. This makes the middle of the laptop too busy.

I suggest you double-click the laptop group, select the keyboard and the touchpad, and give them another color. First, color them the darkest blue we used (the same as the trousers of the sitting guy, for example) and mix a new milder color from that dark blue and the gray of the laptop under it. This will make the character stand out from the background, as seen in *Figure 4.43*.

Figure 4.43 – Using local color mixing to create new colors to enhance contrast

The only character left is the woman sitting on the cloud with her laptop. This is yet another great opportunity to practice adding shadows to a character using the sandwich method:

1. Double-click the group of the character, select the torso, and duplicate it. Move it to the right and lower it a bit this time. Duplicate the torso again and create the intersection of the two duplicates. The intersection will be the light part, so you have to apply a lighter blue for that.

2. The same goes for our legs, as before. Use the **Bezier** tool to draw a path covering part of the legs, then duplicate the leg and create an intersection with the path you drew. Repeat this for both legs. In the case of this character, I used the same color as the other sitting figure, since their trousers are the same color.

Figure 4.44 – Adding shadows to the third character

One more thing to fix about this character: her seat. A cloud should be soft and fluffy, like a comfortable beanbag. So, let's add some more *fluff* to the cloud for better seating.

3. Select the cloud and duplicate it. Scale it slightly bigger and position it down to the left to create a kind of cushion and seat for the character (see *Figure 4.45*).

4. Then duplicate the original cloud again and create an intersection of the duplicates. Color the original cloud shapes a darker blue to see the difference between the two.

A woman in a blue shirt sitting in a blue cloud chair. This scene is nice but does not have enough contrast. It has at one part where her legs are, but not enough contrast with her shirt. If we make the cloud lighter, then there will be no contrast with the frontal seat part we just built. The solution is to apply **gradients**.

> **Tip**
> It is easy to overdo gradients since they are fun and colorful. My advice is to use them only when you need them. They are useful in situations where you need to create simple shadows or create contrast on a surface. Gradients can really add value to your illustration when used with care.

In this case, a gradient from darker blue to lighter would be the solution. The light part at the top will provide enough contrast for the character's shirt. At the same time, the dark end of the gradient will stand out against the light seat of the cloud. The top of the seat will also blend into the light blue background, which will just add to its cloudy feeling. See my version in *Figure 4.45*.

Figure 4.45 – Building the cloud chair and adding a gradient for contrast and depth

There are two bigger surfaces that could benefit from a gradient – the large cloud behind the laptop and the surface of the laptop under the keyboard:

1. Apply a diagonal gradient fill to the cloud, dark blue at the bottom right fading into a slightly lighter blue on the top-left part. This will give a nice dark contrast at the bottom, and still enough contrast for the shirt of the standing character on the top.

2. Also, double-click the big laptop, and apply a horizontal gradient to the rectangle under the laptop's keyboard. It should be gray on the right side and white on the left.

    This will make the whole illustration more engaging since the left corner of the laptop will blend into the background, but the viewer will still recognize the laptop's shape because of the big blue cloud behind it.

    The big laptop is much better this way, as it is creating a nice backdrop for our characters. But it still feels like it is floating, as do the two characters at the front. The third character is sitting in the clouds; she is supposed to be literally floating! To fix this, let's add a few simple shadows under these characters and the laptop.

3. Double-click the laptop group again and select and duplicate the square under the keyboard. Send it to the back, and move it downward a bit to create a drop shadow for the base of the laptop. Instead of the gray-white gradient, color it light gray, which will create a good shadow color on the white background.

4. For the man standing, double-click the group and add a simple flat ellipse as his shadow. Draw the ellipse, color it the same gray as the shadow under the laptop, and send it behind the character with *Page Down*. You don't need to add any details; as this is a flat-style illustration, an ellipse will suffice.

5. Copy the small ellipse and paste it under the feet of the sitting man too. Again, it does not have to be a detailed shadow; it just has to indicate that he is standing on solid ground.

6. Finally, paste the shadow ellipse again, this time under the small circle on the right. This will put this circle onto the same plane as well. Scale it smaller since it is a smaller object than the two characters. The final resultant image will look like *Figure 4.46*.

Figure 4.46 – The final illustration with all its details

The laptop is standing firmly now, casting a nice shadow. And so are the two characters! All three characters have shadows on them. You even added gradients for better lighting. If you look at *Figure 4.39*, you can see the difference that the simple shadows and the lights make to an illustration. By adding a few extra elements, you create space around the objects and a connection between them.

Saving your illustration

The only task left is selecting all the elements in your illustration and creating a group of them. This will help you move and scale them later as one object.

A complex vector illustration like the one you created during this project can work on different platforms. You can use it on a printed brochure, it can appear on a t-shirt, and it can tell a story as part of a website design (as we will create in *Chapter 6*, *Flexible Website Layout Design for Desktop and Mobile with Inkscape*). Even more, it can be animated or taken apart for other interactive uses.

Since there are so many usages, just save it as a standalone .svg file, only holding the final illustration and not the sketches, tests, and duplicates. From this, you will be able to generate anything you might need later on.

Summary

Congratulations, you have created a complex illustration in this chapter! This time, you started with a vector sketch of simple shapes. Then, after creating this base for the composition, you built three unique characters to tell your story! You practiced using the sandwich method to create shapes that perfectly fit together and learned about local color mixing to apply and mix new colors on the spot.

The characters are talking, moving, and working around the huge laptop that you created as a backdrop. Building that was a great opportunity to learn a productivity trick about applying complex elements to surfaces of different angles. And finally, you turned a flat design into a fleshed-out illustration with added shadows and lighting!

Now that you have finished this complex illustration, you will learn more about image editing in Inkscape. While Inkscape is a vector program, photos can be a part of your workflow too! In the next chapter, we will be Edit a Photo and Create a Hero Image in Inkscape!

5

Edit a Photo and Create a Hero Image in Inkscape

Inkscape is vector graphics software, but it has limited tools for working with photos too. You can use these to your advantage while creating a trendy and expressive photo-based tech illustration! This project will focus on combining vector elements with photos and teach you about clipping and masking capabilities in Inkscape. These skills will be useful later when you decide to enhance your vector designs with added photos.

In this chapter, we'll cover the following main topics:

- When to use photos in Inkscape
- Preparing the photo
- Adding a depth of field effect with blur and masking
- Drawing *into* the photo
- Using clipping and masking on the illustration
- Adding lighting to the image

Technical requirements

Download the photo we will work with during this project. Feel free to use any other photo showing a hand with the palm up.

Here is the image you can get for free and use it in this project: `https://github.com/PacktPublishing/Inkscape-by-Example/tree/main/Chapter05`.

When to use photos in Inkscape

Inkscape is a vector tool, and it is common knowledge among designers that vector graphics are not meant for photo editing in the classical sense. But there are some valid cases when the elementary photo editing abilities of Inkscape are proven useful, which are as follows:

- **Case 1**: When you want to combine actual photos and vector illustrations in your design, you might scale them and rotate them on the spot. Photos can enhance your design and they are useful when creating flyers, posters, or publications – things that you might create in Inkscape. You can simply import bitmap images into Inkscape, and scale, transform, or rotate them almost as freely as vector objects.

  There are times when it is easier to scale and transform a bitmap image right there in Inkscape without breaking your workflow. The possibilities are a bit limited, of course – since Inkscape was not made for this. But all the filters you apply and all the masks you create are flexible and easy to change right in Inkscape. You do not have to switch between programs to scale photos, add simple color effects, or apply masks or filters. Even basic retouching can be done on the fly without opening a bitmap photo editor program.

- **Case 2**: Tracing! Using a bitmap image to only trace a part of it is why most beginner users open Inkscape for the first time. You can trace by hand using the **Bezier** tool or use the **Trace Bitmap** function under the **Path** menu. Both are useful and a great way to take photos. After tracing, the original photo can be deleted from the file since it is not planned to be part of the final design.

These are the main cases when you will edit photos in Inkscape, and you will practice a bit of both in the main project of this chapter.

In this project, you will create a hero image, as shown in *Figure 5.1*, a wide banner that can be used as the main header image of a website. You will learn basic photo editing and practice all the tricks you learned while creating the illustration in *Chapter 4, Create Detailed Illustrations with Inkscape.*

Figure 5.1 – The hero image you will create in this chapter

Preparing the photo

At the base of this project is a photograph that you will draw a vector illustration on. You can download the photo used in this chapter from the link provided in the *Technical requirements* section. It is also fine to use another similar photo for this project. If you do so, make sure that the photo you want to use has good resolution and shows someone with an open hand, palm facing up, similar to the original photo.

After you have obtained your image, import it into Inkscape using **File** | **Import**, or *Ctrl + I*.

You can learn more about importing your bitmap images into Inkscape at the beginning of *Chapter 2, Design a Clever Tech Logo with Inkscape*.

The image we'll use is quite straightforward, and the main topic is clearly visible against the background. Yet, we should apply some initial changes to it to make the image appear wider and give more space to the illustration we plan to draw on it.

Clipping and masking

Clipping and **masking** are very similar actions in Inkscape. They are both used to apply a vector mask on another vector shape, group, or even photo (as you will do in the current project). They are similar, but there are some important differences that we need to clarify.

When you are clipping, Inkscape only uses the shape and position of the clipping mask. Color information, opacity, and filters are ignored and do not affect the image in any way. This is why you can create a semi-transparent shape and use it as a clipping mask. It is also very convenient to use any shape with a random color and just create your clip with that.

Masking, on the other hand, works differently, as it also uses the color and opacity of the masking shape! When you apply a mask, Inkscape uses the darkness of the mask shape to set the transparency of the masked object. A white mask with full opacity means zero transparency, while a black mask with 100% opacity causes the final image to be absolutely transparent. Apart from color and opacity, blur is also applied to the masked object.

As you'll see, there are more variables to consider when applying a mask than using a simple clip. Both methods have their uses, and we will practice both of them in this chapter.

Clipping the photo

After adding focus to the photo, let's cut the composition of the image tighter. To achieve this, you will use a simple clipping:

1. Draw a rectangle above the photo and scale it, as shown in *Figure 5.2*. This shape will be the clipping mask. Set the rectangle to semi-transparent, with an opacity of around **50%**. This will help you to see the photo under it, so you can scale and position the clipping rectangle properly.

2. When the rectangle is positioned, select both the rectangle and the photo. Right-click and choose **Set Clip** from the drop-down menu. This will clip the photo into the size and shape of the rectangle you placed above it.

Figure 5.2 – Using a rectangle to clip the photo under it

Your clip is now applied to the photo, and it seems that it is cut to the shape of the clipping rectangle you used. It only seems so because this method preserves the original image! You do not actually *erase* parts of the image but only hide them.

This means if you are not satisfied with the results, you can right-click on the image and select **Release Clip**. This will separate the clip from the photo again. No image data is lost; you can refine the clip and apply it again.

Adding the depth of field effect using a mask

Depth of field, or depth of focus, is how sharp or blurry the subject in front of the camera lens appears compared to its surroundings in the photo. It is a great tool that photographers use to bring their subject into focus. Usually, the subject appears sharp, while the background is blurry, which brings the viewers' attention to the important parts of the image. It is also a widespread digital effect you can recreate using bitmap photo editor programs. Even smartphone photo apps apply an automatic blur filter to create focus in a portrait nowadays.

Using masking, you can create this effect in Inkscape as well! This is not a widespread method in Inkscape, and masking is a bit more complex than clipping but can create more interesting results.

You will draw an illustration over this photo later on, with the main part in the hand of the character. This is where we want to focus the attention of the viewer. The goal is to blur the suit and tie in the background while keeping the hand sharp.

First, you need to remove the clip you added earlier. This happens a lot while using Inkscape, so it is important to practice it. Right-click the clipped photo and select **Release Clip**. This will make your clip appear over the photo again, and the hidden parts of the photo will be visible again.

Of course, you will need to re-apply this clip later. So, select the clip and send it behind the photo with the *End* key! This way, it will be positioned in the exact location as before, just hiding under the photo.

Select the photo and apply some blur to it. Do not overdo it; keep the suit recognizable, but blur some details.

Now duplicate the photo with *Ctrl + D*. Do you remember the sandwich method from *Chapter 4, Create Detailed Illustrations with Inkscape*? We will do something similar, but with masking this time! The duplicated photo is over the original one, in the same position. Select the duplicate and set the blur back to 0.

Now the blurred image is underneath, with the sharp image over it. Select the **Bezier** tool and draw a shape over the hand and forearm of the character. The simple background is very forgiving; as you can see in *Figure 5.3*, you do not have to trace the hand perfectly!

Figure 5.3 – How to draw the mask over the photo

This shape will be the mask you will apply to the sharp photo. To get that right, set the fill color of the shape to white with full opacity, then add a horizontal gradient to it, as shown in *Figure 5.3*. The white part of the mask will be visible, and the transparent part of the mask will be applied with the same opacity as the mask.

Select the sharp photo and the shape over it, right-click on them, then select **Set Mask** from the pop-up menu. If you set the gradient correctly, the sharp hand and forearm will smoothly fade into the blurred arm and background. The same applies to masking as to clipping: if you are not satisfied, right-click the masked photo, select **Release Mask**, and try again! The image information is still there, and nothing has been deleted.

Finally, select the hand part and the blurred photo, and group them with *Ctrl + G*.

Figure 5.4 – The masked sharp photo over the blurred original

Now the photo looks more interesting, and the focus effect works well, but the edge of the photo is also blurred. But there is an easy solution: remember the clip you created earlier? That rectangle is still under the photo!

Select the group of photos and send it to the back with the *End* key. This will make the clipping rectangle appear on top again.

Clips and masks can be applied to photos and shapes and groups as well. Select the group of the photos and the rectangle and apply the clipping rectangle to the group by right-clicking and selecting **Set Clip**. This will seemingly trim the photo back to the shape and size it was before.

Figure 5.5 – The clipped group containing the blurred and the masked photos

The masked hand and the blur will be intact, and the focus effect will not disappear. You can use multiple masks and clips in Inkscape using groups.

> **Tip**
>
> Blurring and working with photos in Inkscape can be taxing on your computer's memory. If you notice any lag, switch to **No Filters** in the **Display Mode** section of the **View** menu. The effect will be intact, but Inkscape will not render it, freeing up memory.

Making the background of the photo wider

Now that the photo is clipped and better focused on the subject, you need to create some blank space. We need this space to provide a smooth background for your illustration. Since the character is turning left, and its hand is gesturing in that direction too, this is where you want to create more space.

If you just click on the photo and resize it, it will stretch, and this is not what you need. To create space, you need to seamlessly expand the background of the photo. You will not actually widen the photo but add a shape to the left to make the image appear wider.

This method also provides a clean enough background for any text you need to place on the image later. The text needs to be read, and a blank and homogenous background adds contrast and helps legibility.

First, draw a rectangle the same height as your clipped photo.

Position this rectangle over the left side of the photo, with some overlap. You will need this overlapping to create a smooth covering. Don't be afraid of covering up some of the hands as well, as shown in *Figure 5.6*.

Figure 5.6 – Position the rectangle on the left with some overlap

After positioning the rectangle, apply a dark blue color to it using the **Color picker** tool. Hold the mouse button and use the local color mixing method we learned about in *Chapter 4, Creating a Detailed Illustration in Inkscape*. This works great on bitmap images because it will get the average color of the selected pixels of the photo.

After coloring the rectangle, use the **Gradient** tool and apply a linear gradient to it. The left side of the gradient should have full alpha, while the right should be transparent. Position the gradient handlers, as shown in *Figure 5.7*. The trick is to cover the edge of the photo: make sure that both gradient handlers are over the photo so the transparent parts of the gradient overlap the photo!

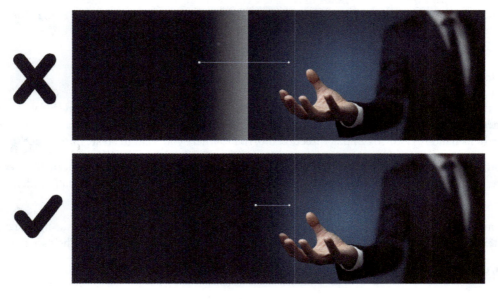

Figure 5.7 – How to apply the gradient to cover the edge of the photo

After fine-tuning the gradient of the rectangle over your photo, select the covering rectangle and the photo group and group them with *Ctrl + G*.

> **Tip**
>
> Covering using shapes with gradients is a handy trick I use a lot to create space on one side of photos in Inkscape as you just did here. It can be a smooth place for text or cover up unwanted parts of the image. And, of course, the same trick works for covering up a sky with light blue colors or adding a smooth background to a busy image. This simple method offers a lot of possibilities.

Using a gradient cover, the photo now appears wider. This has created enough extra space for your illustration and any text you would like to add later when you use this hero image. You have also created focus by adding a blurred version of the photo and using a mask. The photo now looks ready, and the next step is to create the vector illustration you will combine with it.

Drawing into the photo

In this project, we will be drawing *into* the photo instead of drawing onto the photo. It means we will use clipping and masking to make the illustration appear as part of the photo. We will fit the vector drawing into the natural space of the image. You will learn some useful tricks about this method later in this project, but first, you need to create an illustration to use it!

Preparing a simple illustration

A good designer is resourceful – in this case, this means reusing elements you have already created in previous chapters of the book. These are all simple shapes, but it is much easier to copy and paste them than to draw them from scratch. You will need a cloud, a human icon, and a lock icon, as shown in *Figure 5.8*.

Figure 5.8 – Reuse the cloud, human, and lock elements to form a new illustration

First, you will need the cloud you created back in *Chapter 3*, *Modular Icon Set Design with the Power of Vector*. Copy the cloud shape from there into the active work document with the photo. This cloud will serve as the center of the illustration.

Second, copy the human shape we designed back in *Chapter 2, Design a Clever Tech Logo with Inkscape*. This little user icon will be at the nodes of the network of the illustration.

And third, to make things a bit more interesting, copy the lock icon from *Chapter 3, Modular Icon Set Design with the Power of Vector*.

If you skipped the previous chapters or did not design those elements yet, look for the description of those pieces in the mentioned chapters and create them now!

Changing the lock icon

If you copied all three elements into the file, you will see that their style is different. The cloud and the human icons are flat, colored shapes, while the lock is an outline icon. Set the first two aside for now and focus on the lock. That needs to be turned into a flat version too.

To achieve this, first, select the body of the lock and give it a fill color. It can be any color; for now, I have used blue.

Most of the lock icons are built from outline strokes. Select all the parts of the lock and turn all the strokes to paths with **Stroke to Path** in the **Path** menu by pressing *Ctrl + Alt + C*. This will make the shapes easy to color and merge together later.

We need to cut the keyhole in the middle of the lock. Select the circle and the vertical part of the keyhole and merge them into one shape with **Path | Union**.

Then select this new keyhole shape and the body of the lock, using **Path | Difference** (by pressing *Ctrl + -*), and cut the keyhole into the lock shape.

Now select all the parts of the lock and, using **Path | Union** (by pressing *Ctrl + +*), merge them into one single shape.

Figure 5.9 – The steps to turn the lock icon into a flat-style icon

These might seem like extra steps, but changing an existing design, however simple it may be, is usually easier than drawing an icon from zero.

Creating the network illustration

Now you have to create the network illustration over the photo. The illustration is built up of two main parts – the nodes of the network and the lines connecting the nodes. It is smarter to start the network with a layout of the node elements. If you create this layout first, it will be easier to connect the nodes.

Take the cloud shape and position it over the hand in the photo. The main part of the illustration will be the hand holding the cloud. This is why the cloud has to be in the center of the illustration too.

Now position the user icon on the photo, not far from the cloud. Duplicate this icon multiple times and spread these other users around too.

Repeat the same positioning and duplicate the lock icon too! This will make the network more interesting than just a simple grid of nodes. Do not overdo it; find the balance and just use a few copies of these icons!

Finally, draw a small circle, the size a bit bigger than the head of the user icon. Duplicate this circle over the photo to create 8-10 extra nodes for your network. Take a look at *Figure 5.10* to see how to spread out the network nodes.

Select all the nodes, locks, user icons, and the cloud and color them the same light blue.

Figure 5.10 – The cloud and duplicated nodes spread out over the photo

When every node is in position, you only have to connect them to create a network. Using the **Bezier** tool, connect the nodes to the cloud and to each other with straight lines. Use the same light blue color for these lines you used for the nodes, and keep them thin and straight.

Leave the end of the network open, and have some lines leading outwards. These loose ends will hint at a bigger network. You do not have to run the lines precisely till the edge of the photo; you will fix that later. For now, just be playful here; take this as a simple puzzle. There is no rule about how many lines can connect to one node; just take care that wherever the lines cross, there has to be a node.

When you are satisfied with the network you have drawn, select all the lines and nodes and group them with *Ctrl + G*. Leave the cloud out of this group; you will work with that soon.

Figure 5.11 – The network of lines connecting the nodes

It is time to hide those loose ends. As you can see now, they seem to be cut abruptly. You can change that by applying a simple mask to the group containing the network.

Draw a white ellipse covering the network illustration; this will be your mask. If you just apply it to the network now, it will hide all the parts of the group outside of the ellipse but still will create sharp ends. You have already learned about using gradients on masking shapes to create a smooth fading effect.

But this time, add some blur to the ellipse. Inkscape also calculates the masking area using blurred masking shapes. In this case, it is also faster and gives you more control over the area you would like to show.

Figure 5.12 – The position and amount of blur of the masking ellipse over the group

Position, scale, and blur the ellipse, as shown in *Figure 5.12*. When you are satisfied with the area covered, select the ellipse and the network group under it, and right-click to select **Set Mask** from the menu. This will apply the mask to the group, and because of the blur, the lines on the edges will smoothly fade away.

Again, as with clips and masks before, if you are not satisfied with the result, right-click and select **Release Mask** from the pop-up menu after right-clicking the group. This way, you can modify the mask, change the amount of the blur effect, then set the blurred ellipse as a mask again.

Figure 5.13 – The network group with the mask applied

The network and all of its elements are in place, and with the mask applied, they all blend into the image smoothly.

Clipping the cloud

The only thing that seems out of the picture now is the cloud. You have to work on that to create the illusion that the cloud is somehow part of the photo. Instead of editing the photo itself, you will edit this vector element to look like it is in the photo!

To achieve this, you will clip the cloud to make it look like it is behind the fingers and partly covered by them. Move the cloud aside for now and focus on the fingers in the photo. To create the clipping shape for the cloud, first, you have to *trace* the shape of the fingers holding the cloud.

First, trace the fingers using the **Bezier** tool. Start rough with straight lines, then refine the curves with the **Path editor** tool. Use a color that is easy to differentiate from the background – I have used bright green against the orange skin tones of the hand. Set the opacity of the shape to 40-50% to see what the elements under the shape will cover.

Draw the shape of the fingers in the front since these will be in front of the cloud, and ignore the thumb. That will be behind the cloud, so it does not have to be part of the tracing. You can also keep the bottom part flat, as that will not be part of the clip either.

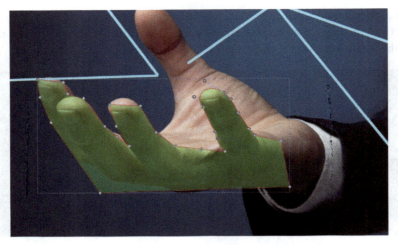

Figure 5.14 – Drawing the clipping shape tracing the fingers

When you finish the final shape for the fingers, put the cloud shape back into its place. But do not apply the clipping shape yet!

Why? Because this is not the final shape. If you set this shape now as a clip, it will only reveal what is under it! This would create the opposite of your goal and show the cloud only where there are fingers.

To avoid this, draw a rectangle covering the cloud. Then, send it behind the shape you drew earlier (using the *Page Down* key), and selecting both, use **Path Difference** to cut the shape of the fingers out of the rectangle. For a better understanding, see *Figure 5.15*.

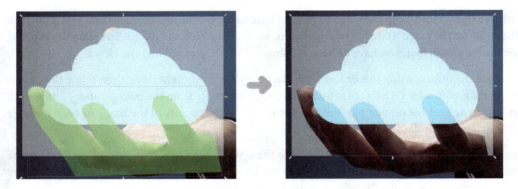

Figure 5.15 – How to create the clip from a square

After you cut the finger shapes out of the rectangle, you have your clipping shape ready. Now you are able to apply this to the cloud so the parts that are supposed to be hidden by the fingers will be hidden.

I explained earlier that clipping the cloud is better than cutting it because this method preserves the original shape and makes editing easier. But clips (and masks) can be applied to groups as well! The advantage of this is that you can move elements inside a clipped group.

In this case, the position of the cloud on the photo is flexible, but the position of the fingers on the photo is given. You cannot change the position of the fingers, but it is beneficial that you can move the cloud *behind* the fingers to decide where it looks the best.

So even if the cloud is one object only, select it and hit *Ctrl + G* to create a group out of it. This is a group now holding one object; the shape you will use for clipping has to be outside this group.

Select the cloud group and the clipping mask over it and choose **Set Clip** from the right-click menu. This will set the clip on the group, and parts of the cloud will disappear (seemingly) behind the fingers.

And here comes the trick: if you want to modify the position, color, or shape of the cloud, just double-click on the clipped group to get inside it. Now you can select the cloud shape and modify its properties.

While you are inside the group, select the cloud, and add a gradient to it. Keep the bottom part light blue and set the top to a darker blue. This will make it look like light is coming from under the cloud. We will add some more lighting later to work with this.

Figure 5.16 – The clipped cloud looks like it is behind the fingers

If you are not satisfied with the clip itself, though, you can select **Release Clip** to modify the shape of the clipping object, then re-apply it by selecting **Set Clip**.

> **Tip: clipping on groups**
>
> Setting a clip or a mask to a group of objects can be really useful! It is great for drawing the eyes of characters, for example, since you can come back anytime and easily change where the eye is looking. Or you can set a clip to a group of objects that you want to show on a mobile screen, modify the elements, and try out different arrangements. This has endless applications, really; practice and use it to make your life easier!
>
> The way we apply clips to groups is not the same as the **Clipping Group** option in the right-click menu! That is an automated function, and it duplicates the selected object and sets this copy of itself as a clip. It is almost the same, but you have less control over the outcome as the shape of the clip is given.

The clipping of the cloud makes it look like it is part of the photo. To complete the illusion, we will add lights to the cloud and network.

Adding lighting to the image

There are two ways you will add lighting to this illustration to make it more interesting. First, you will create the light that is under the cloud. Second, you will add a simple glow effect to the network.

Adding a glow under the cloud

The first method is about creating a shape that will act as the light on the palm of the hand. Since the palm of the hand is lit, this is easy to achieve:

1. Select the **Bezier** tool and trace the bright parts of the palm. Use the same method as earlier when tracing the fingers to clip the cloud shape: start with simple straight lines and refine the shape using the **Path editor** tool. As you can see in *Figure 5.17*, I have created more than one shape to cover all the bright parts on the palm and between the fingers.

2. Select these shapes and combine them with **Union** from the **Path** menu. Do not forget to send the light shape behind the cloud with *Page Down*!

3. Set the fill color of this shape to bright blue and its opacity to 40-50%.

4. Add a horizontal gradient to it, so it softly fades out to the right, around the wrist.

5. Finally, add a small amount of blur to the shape to create an even softer effect. See my version in *Figure 5.17*.

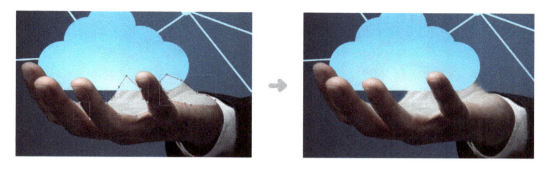

Figure 5.17 – Drawing the lit parts under the cloud and adding gradient and blur

This shine under the cloud seems like a minor change, but this small modification elevates the whole illustration.

Creating the glow around the network

To give even more light to the illustration, let's add a light glow to the network itself. This will not only highlight the lines of the network but give it a digital hologram feeling. We will use a simple method to add the glow around the network:

1. Double-click the network illustration to get inside the masked group.

2. Select all the elements of the network with *Ctrl + A*. If you are inside a group, *Ctrl + A* (meaning select all) only selects the element inside of the given group. So, if you are now working inside the masked group, only the objects and lines building up the network will be selected, and nothing outside the group.

3. Group these selected elements again with *Ctrl + G*. This will create a group inside the masked group.

4. Duplicate this group with *Ctrl + D*. Now you have the network illustration inside the masked group twice.

5. Select the duplicate, and without moving it, apply a small amount of blur to it! This will create the glow effect, as seen in *Figure 5.18*!

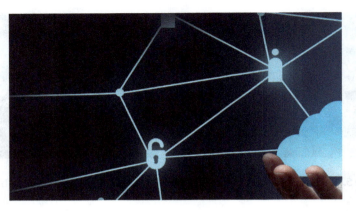

Figure 5.18 – The glow around the network illustration

Why do you have to click inside the group first to achieve this? The reason is to spare time and resources. You could simply duplicate the masked group but doing so would create a hard-to-render illustration for Inkscape.

Apart from using fewer resources, it is also easier to work with a simple masked group containing the network elements. With a few clicks, you can easily modify the elements again if you need, including the masking shape.

Saving and exporting

Now that the image is ready, it is time to save it in the correct format.

If this is to be an illustration only, you might use the image as a `.svg` file on a website. And of course, it is clever to save it as `.svg` for later editing and to easily copy it into other `.svg` files. Save the file as `.svg` to preserve the vector elements.

But if you want to use it straight away, it is better to export it as a bitmap image for the following two reasons:

- **The effects we used**: The added blur effect, masks, and clips may not appear the same way on every platform. Besides, older browsers and computers with fewer resources might struggle with the memory load.

- **The base of the design is a photo**: Using the image as a vector file will only work if the image is embedded into the file or the linked image has a clear path and the `.svg` file can always find it (without this, the photo would not load in the design).

So, if you aim to present your design as it was intended – and of course you do – then export it into a bitmap format. This way it will keep all the effects while sparing memory and loading times. You might choose `.png`, `.jpg`, or the new `.webp` format for online usage.

Summary

The aim of this chapter was to highlight the photo editing capabilities of Inkscape. You started this project by learning more about clipping and masking and clipping the photo to focus on the main subject. Then you added the depth of focus effect, using blur and masking. Then you modified the image via an added blank space that seemingly widened the photo.

During this project, you drew an illustration over the photo. You used the previously created elements and repurposed them. Furthermore, you built the illustration *into* the photo with the clever use of masking and clipping.

Finally, you added lighting and glow to the image to create a modern atmosphere suitable for a tech company. In the next chapter, you will learn how to use Inkscape to create a responsive website template.

6

Flexible Website Layout Design for Desktop and Mobile with Inkscape

Any current website has to be responsive, with resizable elements built on a flexible layout to work great on all devices – and what is a better tool to create flexible layout designs than vector? In the following project, we will create a simple website layout for two different sizes, using the elements we created in the previous chapters.

In this chapter, we are going to cover these main topics:

- Using Inkscape for web design
- Creating a simple wireframe as the base for the layout
- Going from a wireframe to a full design
- Saving and exporting your website design

Using Inkscape for web design

Modern web design has several components, such as user **experience design** (**UX**), content creation, **search engine optimization** (**SEO**), and, most importantly for us designers, graphic design.

Fresh, illustrated websites with unique icons and gradients are everywhere, and oversized type elements and gradients are in trend again. You can create and incorporate typography, illustrations, icons, and many other website elements directly in Inkscape.

And of course, all of these need to look great on all devices from the smallest mobile screens to 32" desktop monitors. All its vector capabilities make Inkscape a great tool to design a responsive layout for any website.

What you will create in this chapter

During this project, you will design a landing page layout for the CloudUsers company. You will first create a simplified wireframe. Then, you will use most of the elements you created so far in the book and create two versions from the site: one for a wider desktop view and one for a small mobile screen.

If creating a website is not your territory, you should still give this chapter a try. It is about practicing your Inkscape workflow, and about seeing what you created so far coming together under a single project. Here is what you will create during this project:

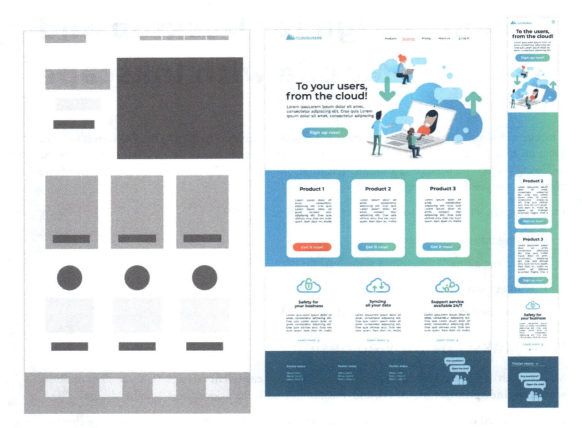

Figure 6.1 – From simple wireframe to desktop and mobile website design

In the following section, we will create a simple wireframe, and then move on to creating the desktop layout version.

Creating a simple wireframe as a base for the layout

A **wireframe** is an overly simplified version of a website. When we say simple wireframe, what we really mean are empty placeholder boxes. There are no colors, no images, and no text to distract you from the first task of a web designer: the arrangement and flow of the information on the page.

There are several apps online to create wireframes and help you sketch your website ideas, but if you already use Inkscape, you might just create your wireframes in vector as well!

Before starting to work on the layout, you need to decide the screen sizes you design for. As you know, we face a wide array of screen sizes nowadays. This forces us to design for most of the brake points in responsive design, ranging from under 320 **pixels (px)** to sizes above 1920 px.

Now, we need to decide in what order things will be presented on the website.

The landing page for CloudUsers will include the following:

- An area for the logo and the navigation
- A header part to introduce the business
- Products to choose from
- Company values and benefits
- A footer with additional info and navigation.

We will create a classic three-column grid. It might be overused, but it provides a solid template to practice your design skills in Inkscape. The navigation and the header will be horizontal rows, while the products and services are organized into three columns of equal width. The footer will be a simple row again.

Let's build a wireframe from top to bottom according to this plan!

Starting the wireframe with the base and the header

1. To start, use the **Rectangle** tool and draw a white rectangle as the background of your design.

   Why draw a rectangle, instead of using the page in Inkscape? Because it is easier to color and resize this rectangle if you need to, and it is a better reference point than the page.

   In this example project, we will use a rectangle that is 1600 px wide and 2000 px tall. This width is comfortable to work with and easy to scale for both laptop and desktop users. Whenever we give exact measurements during creating the desktop layout, we will refer to this base width. Feel free to use the same width or pick another for this practice project.

2. Set the height to 2000 px for now since this rectangle is only here to give a frame to your design, and a website can stretch to any height. You will change the height of the white background rectangle on the fly as you build the wireframe design by adding new rows of placeholder boxes.

Your aim here is not to create a pixel-perfect design. You mostly need to apply a visual balance to your layout and arrange the aforementioned content in a clean and orderly way. To achieve this, you will use guides and alignment in Inkscape.

Now, you need to set the width of the content. If the background width is 1600 px, then you want to set your content to 1300 px wide. This will provide enough margin on both sides of the screen and will have enough white space to arrange your content later.

3. So, draw a rectangle that is 1300 px wide and 120 px tall. Select this new rectangle and the background rectangle and move the smaller rectangle to the top edge of the bigger one and center it. This will help us define the measurements for the width of the content and the header bar as well!

Now, you have to create your first *guides* for this design. Guides are very useful and easy to create in Inkscape. There is more than one way to add guides to your design. You can create them manually by clicking on the rulers on the side or at the top and pulling the guideline out of the ruler onto the drawing board. This way, you can create any number of guides anywhere in your document.

4. However, to automate things and be more precise this time, you will turn the second rectangle into guides. Select the rectangle, then click on **Object** in the top menu, and select **Object to guides**. This will create four guidelines; one on each side of the rectangle that has now disappeared.

Figure 6.2 – The four guides at the top of the page, created from a rectangle

The top one will be aligned with the top of the background rectangle, while the bottom one will define the height of the header bar. The two vertical guides will set the width of the content of your landing page design.

> **Tip**
> **Guides** can be hidden by simply clicking anywhere on the rulers. Click again to show them again. Guides will not show up in your exported .png image, but they are saved in the .svg document and can be used by Inkscape or other design programs later.

Now that the guides are set for the header, you can add placeholders for the logo and the menu. The goal here is not to go into details – just draw a rectangle in the size of the logo.

5. Align this rectangle with the guide on the left since the logo is usually located in the top-left or at the top-center part of a website. We used a 60 px X 280 px gray rectangle.

6. The site navigation will also appear here, in the header bar. Add a rectangle, about 150 px X 30 px. At this stage as a designer, I am not thinking about whether the menu items will be buttons, just text links, or links with a bar to separate them. I simply added this rectangle and colored it a lighter gray than the one for the logo.

7. Duplicate the rectangle three times, and so you get four placeholder menu items.

8. Select all four of them and distribute them using the icon in the **Arrange and Distribute** window that says **Make horizontal gaps between objects equal**. This will create an even line of menu items.

9. Select the menu items again and move them so they are aligned with the guide on the right, as shown in *Figure 6.3*.

Figure 6.3 – Logo and menu placeholders aligned to the guides

The next part of the wireframe you will design is the one with the headline and the hero image. This is setting the tone for the site, and this is the message the visitor first sees when landing on the page.

Setting up placeholders for the content of the website

1. Again, use a rectangle as a placeholder for the hero image. This will be a background image in the final design, so it is OK if it is not totally aligned with the guide on the right. In the example, it is set to 900 px X 740 px.

2. Color this rectangle to a darker gray tone so that you know this is a placeholder for an image. From now on, you can use the same tone to represent *image placeholders* in the wireframe design.

> **Tip**
> UX designers and almost any software use rectangles with crosses to mark the place for images and photos. Even in Inkscape, if you switch to a wireframe view, this is how bitmap images are presented. For this exercise (and when I am actually designing a website in Inkscape), I use simple gray rectangles as image placeholders. It is just faster and more convenient that way. Of course, you can also take notes when you work on the wireframe of a design.

The hero image will take up two-thirds of the width of the page in a desktop view, and it sits on the right-hand side of the screen. The remaining third of the space on the left will be used to add a headline, some short text, and a button.

3. Add a rectangle representing each of these two textboxes. Again, as with the menu items, you are not sure about the exact text, and you are only looking for spacing here.

4. Add another darker rectangle under these – this will stand for the **Call To Action** (**CTA**) button. Your placeholder boxes for the hero image and the headline should look like *Figure 6.4*. It looks very simple now, but you are building a good foundation.

Figure 6.4 – The top of the page with the hero image and the title placeholders added

The next row is for the cards showing the products of the company. **Cards** are a compact form of presenting a subject together with a CTA. They are usually a few lines of text and a link or button arranged in a box format. This is exactly how you will draw them for your wireframe.

5. Draw a tall rectangle – the size of the one in the example is 540 px X 380 px. This will be the base of the card. Draw another, smaller rectangle over the first one (260 px X 60 px). Set the color of this darker – this will become a button or link later on.

6. Select both rectangles and center them. Move the *button* rectangle closer to the base of the other one, as shown in *Figure 6.5*.

7. Group the two rectangles to make a card placeholder. Now, select the group and duplicate it twice. You will have three cards now that you can arrange in an even row. Place the first group at the guard on the left, and one at the guard on the right. Then, select all three of them and distribute them evenly with the *Make gaps between objects equal* icon in the **Arrange and Distribute** window.

This will create a neat row of cards, as shown in *Figure 6.5*.

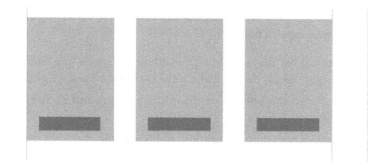

Figure 6.5 – The placeholders for the cards in an even row

The next part is similar, but this row will use cards with icons and will feature some company values. To create a placeholder for icons, designers usually use circles.

8. Create a circle with a diameter of 180 px and color it the same dark gray that you used for the hero image earlier.

9. Draw a small light gray rectangle (340 px X 240 px) under the circle as a placeholder for the text. Then, copy the button placeholder from the product cards and paste it here, under the rectangle. Select the circle and the two rectangles and arrange them by their centerlines. Set a distance between these objects as shown in *Figure 6.6*.

10. Group the elements to make a card and duplicate this group twice.

11. Now that you have all three cards with an icon, do the same as before: place the first one at the guard on the left, the second at the guard on the right, and then select all three of them and distribute them evenly with the *Make gaps between objects equal* icon in the **Arrange and Distribute** window as shown in *Figure 6.6*.

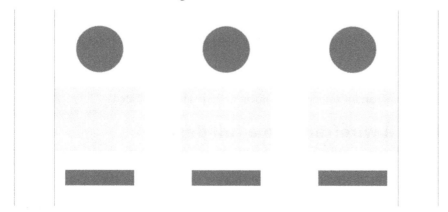

Figure 6.6 – The cards with the icons

Creating an easy placeholder for the footer

The last section of the landing page is the footer. It will be a simple bar sitting at the bottom of the screen and giving the footer menu a home:

1. To create a horizontal bar that matches the width of the *screen* (the white background you drew for your design), simply duplicate the white background and set its height to 300 px.

2. Align this new bar to the bottom of the white background rectangle and make it a medium-gray color.

3. Footers usually have some sort of navigation, additional info about the company, or a simple copyright note. To symbolize this, add a simple light rectangle to this bar, then duplicate it three times, and distribute the four boxes above the footer as shown in *Figure 6.7*:

Figure 6.7 – A simple footer bar with footer menu placeholders

This concludes the wireframe design for our simple landing page. To see the whole wireframe as one unit, check out *Figure 6.1* again, as it is shown in one piece on the left.

The reason to have a wireframe is to set a base for our design and have a blueprint – a place to experiment. If you are not satisfied with a part of the design, you can easily delete it, reposition it, or resize it to fit your needs. It is like a flexible sketch to create before the final design.

Now, you have learned how to use very simple shapes to designate the size and position of the design elements. Then, you added hierarchy and balance using guides and alignment. I hope that this part of the chapter was enough to show you how easy it is to build a wireframe in Inkscape. Even if you don't want to design a website now, you can use this knowledge to create a wireframe for a future project quickly.

In the following pages, we will fill out this wireframe with colors, text, and other visual elements – but before that, let us make an inventory of the assets we have already created!

Going from a wireframe to a full design

A wireframe design is made up of placeholder elements, so it is a placeholder itself. It is only a blueprint of the final version but lacks detail. This is what you will work on in this part of the project.

Normally, when you want to turn a wireframe into a full-fledged design, you need to make a series of design decisions. If you have no company style guide to follow, your task is to build the whole visual style. You need to decide about fonts and colors to use, as well as designing a logo, icons, or even illustrations.

Gathering your existing assets

This seems like a long list, but the good news is, if you followed the previous chapters, and created the projects in this book so far, you already have most of the visual elements for the website!

If you completed most or all of the projects so far, you just need to gather those files. If you have not finished the previous projects yet, maybe it's time to create those elements, as almost all of them will now be used in this chapter.

Here is a list of the elements with a short description and where to find them in the book. As most of the projects are built (in some parts) on the previous ones, it is easier to complete them in this order:

- **Logo design**: The logo is essential for the website, as it sets the visual style you can build on. This includes not just the logo but also choosing the base colors and the fonts you will use on the site. You can create the CloudUsers logo and branding in *Chapter 2, Design a Clever Tech Logo with Inkscape.*

- **Icons**: A set of cloud-themed icons was created in *Chapter 3, Modular Icon Set Design with the Power of Vector.* If you did not create them but want to learn about icons and incorporate them into the website design, then make sure to build at least three different icons from the nine examples created during that project.

- **Main illustration**: You created a complex illustration step by step in *Chapter 4, Create Detailed Illustrations with Inkscape.* All the elements of the illustration represent the company and strengthen the visual appeal of the website.

- **Hero photo**: This one is optional. In the rest of this project, I will show a design based on a colorful illustration. As an alternative to the illustration, you might use the photo-based hero image we created in *Chapter 5, Edit a Photo and Create a Hero Image in Inkscape* instead.

  That will add an interesting style to the website design, and if you prefer that to the illustration, feel free to use that. Using both the colorful illustration and the darker photo-based image would not work well on such a small site, so I suggest choosing one of the two and sticking with it.

It is time to collect all the previously designed components into one document. Let's start with the logo as an example:

1. Open the .svg file containing the final version of the CloudUsers logo. Select the logo where the text is on the side of the cloud emblem and create a group of it using **Object to group**, or press *Ctrl + G.*

2. Select this group and copy and paste it into the .svg document containing your wireframe design.

3. Repeat the same procedure for the main illustration and three of the icons previously created. Having all these elements in the same file might seem too heavy, but since these are vector shapes, the file size and the workload on your computer won't be too much.

You will be able to create your final website design in one document.

Figure 6.8 – All the previously designed components together

These are all the design elements you have created so far, and what you need to continue – but of course, there are other missing parts, such as buttons, bars, and text boxes. You will create those now while assembling the desktop version of the site.

Assembling the desktop version

In the coming pages, all you have to do is follow your wireframe and build the desktop version of the layout. You will put the design components in their final place and add what is missing to finalize the design.

The logo and the menu bar

Starting with the menu bar, you will follow the same top-down order as you did when creating the wireframe. On the left-hand side of the menu bar is the place for the logo:

1. Select the CloudUsers logo you copied into this document and position it over the logo placeholder and scale it to fit the same height. When scaling, take care to maintain the proportions!

2. When you are finished positioning the logo, select and delete the placeholder rectangle.

3. The placeholder for the navigation bar was only a row of simple rectangles too. To add more realistic menu items, add text to the menu.

    The font you used for the logo is a modified version of **Montserrat**, so you need to pick a typeface that goes well with that. For this reason, I chose **Work Sans**, another great sans-serif font. Feel free to use another font – just take good visual pairing and consistency into consideration.

4. Create the first menu item by typing `Products` with the regular weight, **16px Work Sans**. Select the text item and color it to an almost black, very dark blue (`#044b5d`). This will provide great readability on the white background.

5. Select the first menu item and duplicate it. Take this duplicate, position it horizontally next to the first one, and retype the text as `Solutions`. This way, you will not only keep the font type, color, and size of the text but the menu items will also remain on the same baseline.

6. Duplicate and move the menu item again and retype the text as `Pricing`. Repeat the same process twice more to add the **About us** and **Log in** menu items.

The next step would be to arrange the menu items as you did when working on the wireframe version, but since this is a more detailed version of the same site, we have to add two more design elements before that.

First, we have to add a distinction between active-state and normal-state menu items. The rule here is to change as little as possible to signal whether an item is selected or active.

7. Just select the first or second menu item and recolor it to a brighter color. It is clever to use colors that you used earlier in the same design project, so I picked the red color we used in the main illustration (#FF5555).

Since the illustration is already copied into the same file, it is easy to pick this color. This color also pops out against the white background, so it can be used later for the buttons and link colors as well.

8. Besides the brighter text color, you might also add a horizontal stroke of the same color under the active menu item. Simply draw a horizontal line under the text and set the stroke width to 2 px. See my version of the menu in *Figure 6.9*.

The other menu item that needs some special treatment is the **Login** button. It is not really a button, but you have to distinguish it from the other menu items so that registered users can easily find it. Since you copied those previously created icons into this document as well, go and use a small user icon from one of those.

9. Select and copy that simple outlined figure and paste it next to the **Login** text item. If needed, scale it so it is slightly taller than the text. Then, select the icon and the text and group them. This will be important in the next step.

10. Finally, select all the menu items, including the group with the small user icon, and arrange the menu items so that they have equal horizontal gaps between them.

This concludes the navigation menu for now, as shown in *Figure 6.9*!

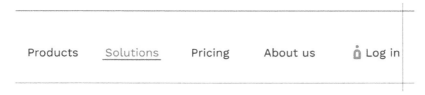

Figure 6.9 – The desktop version of the navigation menu items

Header text and main illustration

After the navigation bar is set, keep following your wireframe layout further down. The next stop is the header text and the main illustration. As you know, both of these items have a similar role – to tell the story of the company and get the attention of the visitor!

In the wireframe, you allocated one-third of the space to the text on the left. Let's start with that. The Montserrat font you used in the logo is also a great display font, so you can use a semi-bold variant of that for the title text:

1. Set the color to the same dark blue you used for the menu before (#044b5d) and the font size to 72 px. This will give a nice block of text where a title or slogan can go. As an example, you can use **To your users, from the cloud!**. Use the placeholder rectangle as a blueprint and delete it after the text is in place.

2. Under the title, create a textbox with the **Text** tool, and fill it with 3 to 4 lines of lorem ipsum, or any other filler text. Set it to the same text color as the other text (#044b5d), but this time, use **Work Sans** with a size of 24 px and a normal weight. This will be smaller and thinner than the title, but still big and easy to read. See *Figure 6.10* as a reference for the text placement!

3. Under the short text, you need to add a CTA button. To draw a button, you need to create a base and add text to it.

    The base of the button should be a wide rectangle – in the example, it is set to 72 px X 320 px.

4. Select the rectangle, and round its corners with the **Rectangle** tool. Set the corner radius to the max, meaning that it will look like a long capsule shape. This way, it will have no corners at all, and it will fit the friendly tone of the website.

5. The button base should have a fresh gradient instead of a flat color. Select the base using the **Gradient** tool and draw a horizontal gradient. Assign the same gradient that is used in the logo – from blue to turquoise! This way, the button will be one of the items carrying the brand colors.

6. Next, add the text to the button. You can use **Sign up now!** or any other phrase you see fit. Set the font to **Montserrat** again, the weight to **bold**, and the size to 28 px, and color it white. This will be thick enough to create contrast against the colorful base of the button. See *Figure 6.10* to see the example.

    Now that the title, the short text, and the button are set to the left, you should move your attention to the right. Here is the placeholder for the main illustration.

7. Since you created the illustration already, this will be one of the easiest tasks while creating the website design: just move the illustration over the placeholder, and scale it to fit into this space. Select and delete the placeholder, and you are done!

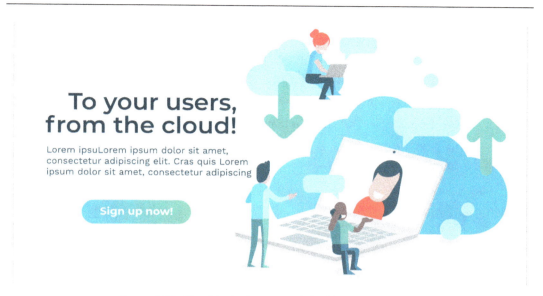

Figure 6.10 – The title the button and the illustration in place

The big hero text and the illustration stand out very well and are a good way to greet visitors to the site!

Product cards row

This section will hold the product cards. You will create the cards the exact same way as you did during the wireframing phase but, of course, this time with added colors, buttons, and text fields:

1. Select the first placeholder card and add 30 px rounded corners to the base rectangle of it. You can do this by setting the **Rx** or **Ry** values to 30 px in the **Tool** menu of the **Rectangle** tool at the top.

   The background color of the cards will be different but leave them gray just for now so the card will be visible on the white background. You will deal with that later.

2. Next, add the textboxes to the card. The title of the card should stand out more than the description of the products, so set the title to **Montserrat bold** again. Use the same dark blue text color as before and set the size of this title text to 28 px. See *Figure 6.11* for more details on the card design!

3. Now, to add the second block of text, scroll back up to the title part, and copy the text object with the filler text from there. Paste the text here into the card and set the width and height of the textbox with the handle at the corner of the textbox while using the **Text** tool, or the **Node** tool!

You can also use the **Flow into Frame** option from the **Text** menu at the top:

I.   Draw a rectangle in the size and position you would like to flow your text into.

II.   Select the text object and the rectangle.

III.   Select **Text | Flow into frame** (or use the *Alt + W* hotkey).

IV.   Delete the rectangle. That's it – it is that simple!

You can check this method in *Figure 6.11*. As you see, you can create precisely sized textboxes with this solution.

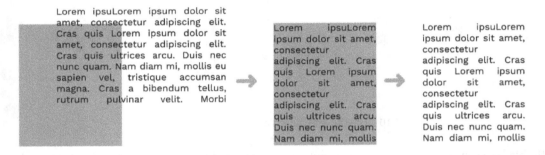

Figure 6.11 – Flow into Frame to precisely resize a textbox

You also need to change the size of the text and set it to a smaller size – it is 18 px in the example design.

Why use a copy of the text object? You could achieve the same thing by drawing a new text object, copying text into it, and setting all the text properties as before, but copying and modifying an existing object in Inkscape is always easier and faster than setting all the values over and over.

4.   The product card has a title and description now, but it is missing a CTA button. Select the button from under the title you created earlier, copy it, and paste it here into the card.

5.   Of course, the button will be too big for this card – you need to scale it down to fit. Set **height** to 60 px and size to 28 px.

6.   Select all the elements of the card and create a group of them. Check out *Figure 6.12* for the final card layout!

The card is almost looking good now, but the background of it is still gray. You want to have it crisp white – this is better for the text and the buttons too. Select the background square of the card and set it to white.

However, now you need to separate the white card from the white background. You could use a thin stroke, or put a drop shadow under the card, but I think it is better for the whole design if you add contrast to the cards by adding a colorful bar under them.

Using a colorful background for the product row is also a good way to separate this section from the previous one. In the title section you just finished earlier, the colorful illustration and the bold, dark text create a nice contrast to the crisp, white background.

You need to then add more color to separate this section from the previous one. The separation of rows helps to guide the visitor through the different contents of the page.

7. To achieve this, first, select and duplicate the white background rectangle that you drew earlier for the website design. Duplicating makes sure that it is the same width and in the same position as the original background. Now, set the height of it leaving some margin above and under the card. In the example, it is 700 px.

8. Send the rectangle behind the card using the *Page Down* key. It is still white, but you will change that soon, using the same method you used earlier when assigning an existing gradient to the button.

9. Select the rectangle and, using the **Gradient** tool, create a horizontal gradient on it. Assign the same gradient you used for the button and the logo earlier. This will strengthen the brand colors and will give a colorful background to the white card!

10. In the wireframe, there was more than one card, of course. This problem is easy to fix – simply duplicate the finished card twice and position the cards where the card placeholders are. Spread them out evenly using the **Distribute** dialog window as before. Once done, delete the placeholders – you will not need them anymore.

You can also demonstrate at this point how the hover state of a button would look.

11. Select one of the buttons from the buttons on the three cards and recolor its background. Use the same vivid red you used while coloring the active menu item in the top navigation bar.

Using the gradient as a vivid background, your cards will pop! See *Figure 6.12* to check out this design solution.

Figure 6.12 – White product cards on a colorful background

This section was good practice to create cards and add a simple but colorful component to the page. Next, you will create something similar but still different for a different result.

Company values with icons

Right under the product cards, we can show a bit more about the company. You can call this part a list of services, or company values – the point is that visually, these are another version of the product cards. Not so direct, but still selling the idea of the company to the visitors.

You drew the layout for this section as a wireframe, and now you just have to follow that blueprint again. There will be three blocks, each containing an icon, a short title and text under that, and a CTA at the bottom.

As these text boxes are almost like product cards, you can work smart and salvage some parts from those cards.

So, to start this section, execute the following steps:

1. Copy the title and the text block from one of the cards you created in the *Product cards row* section. You will not need a card base this time, so just paste the title and the text where it should go according to the blueprint.

2. Modify the title to something else (see *Figure 6.13* for an example) and make the font size of the title a bit smaller. In the example, I used 22 px, instead of 28 px for the cards.

3. Keep the font size of the text box the same but make the text itself a bit wider. There will be no visual card base now, so you will not have to accommodate any margin this time.

4. Next, you need to add the icon to the top of these text blocks. Get one of those icons you copied here from your project in *Chapter 4, Create Detailed Illustrations with Inkscape*, and position it above the title as shown in *Figure 6.13*.

    You will have to scale the icon too. When scaling, take care that you scale the stroke width proportionally too! In the chapter where you designed the icons, you kept the icon for this function turned off on purpose. Now, you need to turn it on so that the lines are all scaled, keeping their width and proportions (if you don't still remember this part, don't worry – go check out the tips in *Chapter 4, Create Detailed Illustrations with Inkscape*).

    At the bottom of these text boxes, you need to add a CTA. This time, you can keep that a bit more subtle – so no button, just a text link with an arrow.

5. Duplicate the title of the box, move it here under the text, and set the text size to 18 px, while keeping the **bold Montserrat** font. Make the text the same turquoise as the icons (#2AC1B5) and retype something such as **Learn more!**.

6. This is a text-only button, but you can still make it a bit more interesting by adding an arrow next to it! Copy a small arrow from one of the icons you created earlier in *Chapter 4, Create Detailed Illustrations with Inkscape*, turn it so it is pointing right, and then position and scale it next to the text. If you need more help, check out *Figure 6.13*!

7. Select the text and the arrow and make them a group. Now, select all the elements of the text box – the icon on the top, the title, the text, and the text button – and align them with *Center on vertical axis* in the **Align and Distribute** dialog. With all of the elements still selected, create a group of them.

8. Duplicate the group twice, and spread them out evenly as before, using the icon in the **Distribute** window to make the gaps between the text boxes equal.

9. The last step for this section is to change the titles in the second and third boxes, as well as change the icons in them. Move the new icon inside the group and position it over the old one. Luckily, we created all icons as the same size, so it is easy to fit them into the same space. Delete the previous one and you are done!

Safety for your business

Lorem ipsuLorem ipsum dolor sit amet, consectetur adipiscing elit. Cras quis Lorem ipsum dolor sit amet, consectetur adipiscing elit. Cras quis ultrices arcu. Duis nec nunc quam. Nam diam mi, mollis

Learn more >

Syncing all your data

Lorem ipsuLorem ipsum dolor sit amet, consectetur adipiscing elit. Cras quis Lorem ipsum dolor sit amet, consectetur adipiscing elit. Cras quis ultrices arcu. Duis nec nunc quam. Nam diam mi, mollis

Learn more >

Support service available 24/7

Lorem ipsuLorem ipsum dolor sit amet, consectetur adipiscing elit. Cras quis Lorem ipsum dolor sit amet, consectetur adipiscing elit. Cras quis ultrices arcu. Duis nec nunc quam. Nam diam mi, mollis

Learn more >

Figure 6.13 – The service boxes with the icons and simple call to actions

The services are now presented differently and interestingly under the product cards section. You can add more of these columns too, arranging a second or even a third row under the original one. For this example, we only created three text boxes with added icons, but feel free to experiment and practice!

There is only the final section left in your website design – the footer and the footer navigation menu.

The footer bar and the footer menu

The **footer** is an important part of any website. It has to be simple and yet contain important data. If the header with the hero image and the big title was the party end, then the footer is the plain business end, but of course, that does not mean that you cannot add a bit of fun there too:

1. To create the footer bar, duplicate the white background rectangle, and set its height to 350 px. This will be enough to accommodate a simplified menu. The color of the footer is usually gray, or another plain color compared to the colors of a website.

   In this case, since the main colors you used for the CloudUsers identity were blue and turquoise, I would choose a pale version of those. In the example, I used the blue color (#0a85a4) that I used in some of the dark blue parts of the illustration created earlier.

   This is dark enough for the background and desaturated compared to the vivid colors of the illustration and the site in general.

2. Now that the bar is set, you will add four columns to the footer. Three of these will be for the footer menu, and the last one a chat button with some extra playfulness.

3. To create the design for the footer menu, simply create a few lines of placeholder text. In the example, I set the font to **Work sans**, normal weight, and a size of 16 px. This will be the smallest font on the site.

4. To make it look even more mundane, set the color to the light blue (#9EECFF) that you used in the illustration. This will lessen the contrast between the footer background and the text. This text is not there to pop – on the contrary, it is only there to be found if someone is looking for it.

5. Set the first line to bold and a slightly bigger font size. That will be the title of the menu list.

6. Now, duplicate this list twice and distribute the three columns as you did before with the cards and other elements in this chapter. You can check the example in *Figure 6.14*.

   The last column in the footer will be a contact link (that is, to a live chat or call). This could be a simple text link, but it is more interesting to create a contact icon here. Of course, you will use parts you have already designed!

7. Copy the graphical part of the logo – the little users in the cloud – and paste it here into the footer. Color it the same light blue (#9EECFF) you used for the footer text.

8. Then, take one of the speech bubbles you drew for the illustration and copy and paste it here too! If it is not the same blue color as the rest of the icons, set it to the same light blue! Duplicate the speech bubble and flip it horizontally (with the *H* key). Position the speech bubbles above the logo as shown in *Figure 6.14*!

   To finish the contact icon, add some text to the speech bubbles. Use the same font as for the footer titles and set the text to the same blue as the footer background. This way, you have created a monochrome illustration that is playful but blends into the footer.

9.   Select all the parts of the contact icon and create a group with *Ctrl + G*.

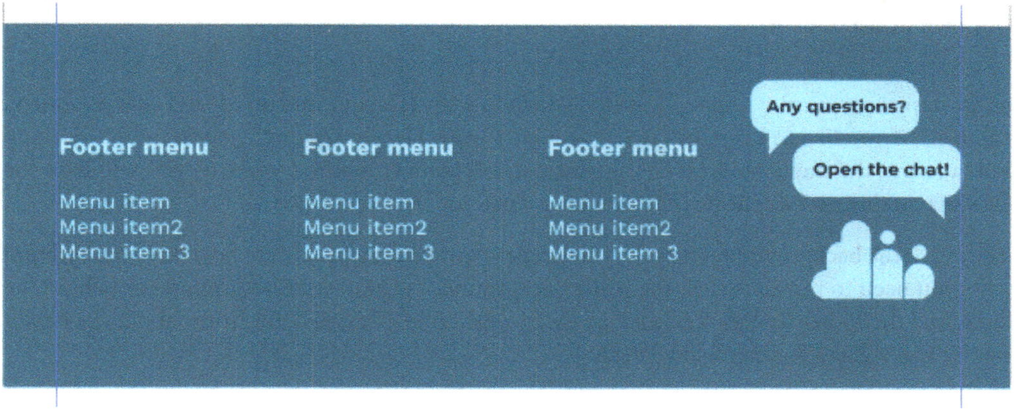

Figure 6.14 – The footer menu and the contact icon

With the footer done, the last part of the website design is completed! If you followed the steps in this chapter so far, you finished a wireframe design and a fully colored website design.

Next, you will create the mobile website based on the desktop version – faster than you think!

Creating the mobile version of the design

This book is all about practicing your Inkscape skills and working smartly and efficiently. This project is particularly about using the components you created in earlier chapters – and this upcoming segment is all about mutating the desktop version of the site into a simple mobile view.

Resizing for mobile

It is a bit of an oversimplification, but the main difference between the desktop and the mobile version of a website is the size. The mobile phone as a medium is narrower and smaller than a computer screen. This creates different usages and usability challenges if you want to provide the same (or a similar) experience to users on mobile. You as a designer have to maintain readability and usability on this smaller screen.

You will not create yet another wireframe layout but instead, modify the desktop version straight. Vector design is always praised for how easy it is to scale things up. Now, you will see how flexible it is to scale things down!

As we established at the beginning of this chapter, you will only create two versions of this website layout: a larger desktop view and a small mobile view. The mobile version will be designed for a 360 pixel-wide screen, so you will have to rescale and fit every component to this size.

> **Reminder**
> The upcoming pixel values are used by me in the example; you are free to experiment and use your own measurements.

To use the design you already created, select the desktop website layout, and save it as a new document. Alternatively, if you like to keep things in one bigger work file, simply duplicate the whole desktop version, and put the duplicate next to the original. Whichever method you use, you will base the mobile view on this new duplicate of the desktop version.

Let's start with the background rectangles – rectangles plural because there are now three background elements you need to rescale. Select the white background, the gradient background bar behind the products, and the footer background and set their width to 360 px. Do not worry about the height for now, as you will set that manually later.

Logo and mobile navigation

After resizing the background, following the same method as earlier, you will resize the rest of the components from top to bottom. There will be some additional elements, but you will work on those later. Now, just resize what you have at hand:

1. First, the logo. Set it to 150 px wide proportionally.

2. Delete the navigation menu – we will hide the menu items behind a simple hamburger menu on the live page. This is an easy icon to create: draw a horizontal bar, that is 4 px X 24 px.

3. Duplicate it twice and order the three bars under each other. This is your hamburger menu icon – group the bars and color them the same turquoise you used on the site (#18B7C2).

   Since this is just a layout plan, we will not show the open menu. See the position of the logo and the menu icon in *Figure 6.15*.

Header and the hero image

Under the logo is the header section with the title text, the button, and the hero illustration. On mobile, you should always decide what is the best way to present text and visual elements. Most of the time, to keep things readable and visible, you need to scale them to fit the screen and arrange them under each other.

This is the obvious reason why mobile sites seem longer and narrower than on desktops. This is exactly what you need to do with the header section:

1. Resize the title text to 36 px and center the text to the background. Otherwise, keep the color and the font the same.

2. Under the title, set the font size of the text to 16 px. Then, resize the textbox with the **Node** editor tool or the **Text** tool so that it is narrower but is split over more lines for the same amount of text. You can also use the **Flow into Frame** method as was explained earlier.

3. The original size of the button is, of course, too big to be used here, but before shrinking it too much, think about usability: buttons on a mobile screen have to be ergonomically built since users need to tap them with their fingers. With this in mind, select the base of the button, and resize it, so the height is 55 px.

4. Then, resize the button text too – set it to 24 px. This will be a big enough button to read and tap but will not overwhelm the design. Select the title, the textbox, and the button, and align them to the centerline of the background.

5. Under this button is the hero image. Scale it down proportionally until its width fits the background with some margin. In the example, it is 320 px wide. Select the illustration and align it horizontally to the center of the background.

With these simple steps, all the elements add together logically and start to form a smaller layout. Refer to *Figure 6.15* to see the full picture so far.

Figure 6.15 – The first sections of the mobile website layout resized

These are the first sections of the website resized to a mobile view. Next, you will work on the products and services and add more mobile UI elements.

Realigning and resizing the product cards

Following down the road again comes the next task: resizing and rearranging the product cards. The cards need to be scaled down while the text needs to be readable. Luckily, the cards are designed in a way that resembles a vertical mobile screen, so you will not have much work to do. You need to resize the cards to fit the screen width and rearrange them under each other so that users can scroll through them easily.

You could apply changes to all three cards, but since this is only a website mockup with no real content, it is easier to change one card and duplicate it twice:

1. To start, delete two of the cards and leave only one to work with.

2. Since you design for the smallest average phone screen, you need to scale down the card background. To fit a screen that is 360 px wide, set the width of the card base to 260 px, and the height to 400 px.

3. Then, resize the text as well: the title can be 28 px and the running text 16 px. Using the **Text** tool or the **Node** editor tool, resize the block of the text manually so that it has more lines and is narrower to fit into the card base. Alternatively, use the **Flow into Frame** option as explained earlier.

   The button has to be resized as well. If you set the height of the base to 45 px, and the font size to 20 px, it will be still readable and usable. Again, these numbers are just guides for this example – experiment with your sizes!

   As you can see in *Figure 6.16*, the visual of the product cards did not change much, but to fit them on the screen, you need to arrange them under each other.

Figure 6.16 – The product cards ordered under each other in the mobile view

4. Create two duplications of the card you modified and arrange them under each other. Select all three cards and center them on the vertical axis of the background.

5. With the cards still selected, distribute them with the **Make vertical gaps between objects equal** button in the **Arrange and Distribute** window.

6. Now that the cards are completed and organized, you need to fit the height of the gradient bar to accommodate them. Select the gradient background rectangle and set the height manually until it has a bit of margin under and above the first and last cards of the column.

The product cards are now rearranged, resized, and presented in a classic vertical manner. In the next section, you will try to visualize another type of user interaction.

Reworking the company values section

Our generation is used to scrolling endless pages on our mobiles, so simply arranging content in a single column works, but you may want to use another type of interaction from time to time to present the information differently. This time, you will order the components next to each other in a row, and only show one at a time on screen:

1. Although different, you will start to reshape this section the same way as the previous one. Delete two of the columns and leave only one textbox with the icon on the top and the text link at the bottom.

2. Again, you will have to resize the text. Set the font size of the title line to 22 px and the font size of the text in the textbox to 16 px. The text link at the bottom needs to be smaller too – 16 px is enough. Shrink the arrow as well.

3. Again, reset the width of the textbox too – make it fit into the screen. Use the **Flow into Frame** method explained earlier or resize the textbox by hand using the **Node** editor tool or the **Text** tool.

4. Next, you need to resize the icon at the top too. Although it fits the screen perfectly, it needs to be smaller. It was designed to be small and simple, and, of course, cannot be as large as the hero image at the top of the page. Resize it proportionally to be 100 px wide.

   Finally, add pagination under the box. Users are familiar with this type of interaction; they will know that there will be more content after they swipe left.

5. To draw a simple horizontal **dot pagination**, draw a 10 px circle, and duplicate it twice. Spread the three circles out evenly with a gap of a few pixels between them. Color the first circle turquoise (#2AC1B5), and the other two light gray (#E6E6E6). This will show which circle – and which textbox – is active.

   As shown in *Figure 6.17*, using dot pagination is an easy but elegant solution to simplify the layout for mobile view.

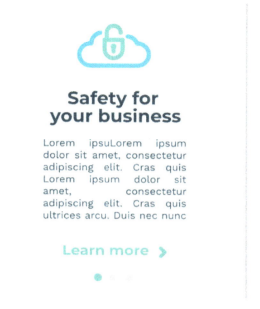

Figure 6.17 – The company values textbox with the horizontal dot pagination

Of course, this is just a static visual design – if you want to see this method in motion, you need to build a simple prototype, and that is not part of this book.

The next part is the last segment – the footer and the footer menu.

The footer menu on mobile

The **footer** is an essential part of a web page, and it was established on the desktop version of this website design. It has to be there in the mobile version too, but since we cannot allocate so much space to it, we have to simplify it. We will keep the footer bar visible, of course, but hide the footer navigation behind a drop-down button:

1. First, delete the menu lists, and keep only one title saying **Footer menu**. You already resized the footer background square earlier, so you only need to fix the text size here. Set the text size to 16 px.

2. Now, select the right arrow that you added to the text link just above the footer in the company values section. Copy and paste that arrow here, next to the footer menu title. Set the color of the arrow to the same light blue as the text in the footer (#9EECFF).

3. Rotate the arrow to make it point downward. This is an indicator that there is a drop-down menu here. To emphasize this drop-down link more, draw a 2 px thick horizontal line under it in the same light blue color.

4. The line is also great to separate the link from the contact icon you have now at the very bottom of the page. Resize that icon and align it to the centerline of the page.

That small contact icon was only sitting at the bottom of the page in the desktop view, but as you can see in *Figure 6.18*, it inhabits the space in the footer in the mobile view.

Figure 6.18 – The mobile footer with the hidden navigation and the contact icon

With the footer also reworked, you have completed the mobile version of the website design. As you saw, this was all just a case of resizing and repositioning elements you had already created. Inkscape is great for working with these types of responsive design elements. All the components are easy to move and repurpose when they are based on vectors.

Saving and exporting your website design

After creating the two layout versions of a website, you usually want to use the design. This means you might present it to a client, export the components and hand it over to a developer, or prepare to build the live version yourself.

Exporting the designs as mockups for a presentation is pretty straightforward. Select the website design and export it as a .png image in standard 96 DPI resolution. This is usually sufficient to give a preview to anyone and show off the work you created.

Saving individual components

To use the designs for building the website, you need to decide what the components that you actually need to export are, and which can be expressed in code. You will need to export the logo, the hero illustration, the icons, and the contact icon in the footer.

Select these components one by one to export them as .svg files. Here is a short description of how to save parts of a website design as .svg files for online use:

1. Select the component you would like to use and turn it into a group.

2. Create a new document and copy and paste the group into it. Make sure that it is a clean document with nothing else in it.

3. Select the group and hit *Shift + Ctrl + R* or go to **File | Document properties | Resize page to drawing or selection** (using the hotkey is faster this time). This will resize the page to the drawing.

4. Now, if you save the document, it will create a .svg file with only the component in it, perfectly sized and positioned. Take care not to save it as an Inkscape .svg – rather, use the **Optimized SVG** or **Plain SVG** formats! These formats exclude all the macros and extra data Inkscape is using in the native files, so you will have a smaller file size and a format that will render the same in every browser.

5. Repeat this process for all the components. Do not forget to also save a .png file with transparent backgrounds for each component. This will be your bitmap image, the fallback file format for browsers or frameworks that are not compatible with .svg files.

When rebuilding a design into a website, exporting the images is only half of the work. The other half is to collect the color codes, measurements, and fonts to be used on the websites.

Take the gradient bar behind the product cards, for example. You won't have to export the gradient as a bitmap image, not even as .svg. You simply check the color code and the angle of the gradient – this time, it was flat horizontal – and use that data in the CSS file to generate the gradient in the browser.

Buttons are not different either. You can apply rounded corners to any object via CSS and even specify the corner radius. Background colors and gradients should come from your original website design as well.

The same is true for the fonts used: use Google font services, or any other method to include the fonts in your website. As it was stated earlier, all sizes in the design are just approximations – you will have to find your own sizes.

As you can see, preparing a website you designed in Inkscape is a simple task if you have Inkscape open. You can easily export elements and get measurements and color codes to use in a real-life project.

Summary

During this chapter, you created three different versions of a landing page design. First, you built a simplified wireframe version. This was useful to set the sizes of the page and figure out the place for the different elements. Then, you collected all the materials you could use from the previous chapters: the logo design, the icon set, and the illustration.

You used most of the components you created so far in the book and practiced resizing and rearranging them into a bigger, more complex design. This is how you built the desktop version of the website – then, after more resizing and rearranging, you turned that into the mobile version. This flexibility is the reason Inkscape works so well for web design! You can rescale and reuse elements creatively and effectively.

In the next chapter, you will learn about optimizing your workflow by combining Inkscape with other applications.

7
Combine Inkscape and Other (Free) Programs in the Design Workflow

A designer rarely uses one software only. Most of us have our own arsenal of different applications for different tasks. Inkscape is great for vector graphic tasks, but it does not have to stop there. You can always use Inkscape in combination with other programs to create a workflow that suits your needs. In this chapter, we will get to know a few ways Inkscape can complement other design programs.

In this chapter, we are going to cover the following main topics:

- Fitting Inkscape into your workflow
- Enhancing your vector art with raster brushes and textures using Inkscape and Krita
- Desktop publishing with crisp vector elements using Inkscape and Scribus
- Turning your 2D work into 3D using Inkscape and Blender

This chapter will be different than the previous ones, as we cannot go into detail about using other apps in this book. Instead, you will get three bite-sized example projects about using your Inkscape creations with other design programs.

During the projects, we will mostly focus on two things:

- The process of transferring your design from Inkscape to other programs
- How the vector art can be used in a relevant way within the software at hand

Feel free to pick any of these three projects, download the required applications, and start working.

Technical requirements

Download and install the latest stable versions of Krita, Blender, and Scribus. You do not need all of them, of course; install the one that you want to try in the following projects:

- **Download Krita**: `https://krita.org/en/download/krita-desktop/`
- **Download Blender**: `https://www.blender.org/download/`
- **Download Scribus**: `https://www.scribus.net/downloads/stable-branch/`

Fitting Inkscape into your workflow

As proven in all the previous chapters, you might use Inkscape in various projects. Various projects – but all vector based. In this chapter, you will broaden the range of possibilities.

Simply put, if vector images can be used in your workflow, then Inkscape can be useful for you. Combined with other software, it can be a worthy tool for anything from raster illustrations to creating 3D models. The use of it only depends on what your goals are and how well you can fit Inkscape into your design workflow.

As a designer, it is important – and inevitable – to learn and use more programs. Do not stick to one – be flexible and find the tools you need. The goals are to be comfortable and find your own working method.

Well, take me for example. Of course, my favorite tool is Inkscape. I use it for everything from logos to game art and I use it almost every day. But even I use different art applications to reach different goals. I am not afraid to combine Inkscape with other programs when I need it, and you shouldn't be either. This is not a competition; use any application you like and create your own design workflow.

Every workflow is different, but commonly, they start with a design brief or an idea. This is followed by sketching and then a design phase until the final product is created. Inkscape is usually important in the actual design part of the process, and mostly *the start* of that.

It is understandable, since creating clean vector elements, and then using them as the base of raster illustrations, or 3D objects is easier than the other way around. That is, creating a digital painting and turning it into a vector works okay, but not perfectly. If your workflow can be built on vector images, you might use Inkscape there.

Transferring projects between Inkscape and other design apps is mainly possible due to all the file formats Inkscape can work with. When you want to start with a vector file, the native SVG format is perfect for most other programs, and if not, you might use PDF or EPS.

It is also possible to create a fast design in Inkscape and export it into PNG format before manipulating it in raster-based software such as Krita or Photoshop. We will learn more about file formats in *Chapter 8, Pro Tips and Tricks for Inkscapers*.

If working in vector can make part of the process faster or easier, then do it. If the quality of your design work can benefit from using Inkscape, use it!

Inkscape and Krita – enhancing your vector art with raster brushes and textures

Inkscape illustrations are usually clean and simple and lack the hand-drawn feeling or texture unless the artist is using hundreds of objects or applying filter effects. Krita is a great digital drawing tool, and it has a whole arsenal of brushes and textures. If we combine these two applications, we can have the best of both worlds!

The clean vector illustrations from Inkscape create a perfect base for the raster brushes and effects provided by Krita! Drawing over these simple vector shapes in Krita can create softer surfaces and add a natural look to your illustration.

> **What will you create in this project?**
>
> You will take an existing vector illustration from Inkscape – about the girl sitting on the cloud with her laptop – and import it into Krita. From there, you will add texture, shadows, and highlights to it using raster brushes.

Preparing your work with Inkscape

Just to make things more efficient, let's use a vector illustration you created earlier. Open the SVG file that contains the complex illustration you created during *Chapter 4, Create Detailed Illustrations with Inkscape*. Do you remember the girl with the laptop, sitting on the cloud? We will use that part of the illustration for this exercise.

To start, get the vector illustration ready to use. Select the girl sitting on the cloud, and group the girl, the cloud, and the speech bubble elements. Leave the green arrow out for now since that was part of the big illustration; here, it would be out of the composition of the image.

If your image has the added shadows and highlights you created in Inkscape, just delete those. You will only need the base shapes of the *cloud girl* now, as shown in *Figure 7.1*:

Figure 7.1 – The base of this project will be this part of the illustration

If you didn't create the illustration earlier, you can create it now, or use any other simple vector illustration as practice. Mind you, in this project, we will use that particular illustration, and following this process might be harder with another illustration.

You don't need to export the image this time

With almost any other program, at this stage, we would focus on how to save and export your vector file from Inkscape, and how to import it into the other program. But not with Krita. Since Krita 4.0, the developers have worked on adding better vector tools to the software, and now, a few versions later, you can simply copy and paste SVG objects into your Krita artboard! This will make the workflow smoother as you will be able to switch between Inkscape and Krita with ease.

But now, just leave the Inkscape document open; you will come back to it soon. Before pasting your illustration into Krita, let's open that program and set up a new document to work with!

Enhancing your illustration with Krita

Size does not matter in Inkscape most of the time, because you can resize vector elements seamlessly, without quality loss. But since Krita is a bitmap program, it is important to set image size and resolution while creating a new file.

Creating a new document in Krita

To start designing in Krita, let's make a new document. If we want to create and print a nice illustration, we are looking for a bigger image size. Let's set the size to **180** mm by **180** mm with **300** dpi resolution. Leave the color settings set to RGB and 8-bit for now:

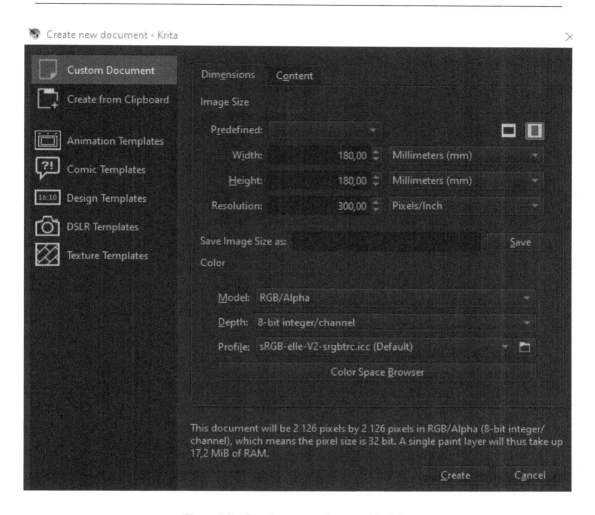

Figure 7.2 – Creating a new document in Krita

Click on **Create** to make the new Krita file. This document will be the basis for your vector illustration to be pasted into.

Pasting and resizing your SVG illustration in Krita

First, you will need the vector drawing from Inkscape. Select the Inkscape window, locate the group with the woman sitting on the cloud, and copy it. Then, simply navigate back to Krita and paste the SVG into the blank document you just created.

After pasting the vector illustration into Krita, you might wish to close Inkscape to save resources.

Notice how the vector image was pasted onto a new layer. You can see the layers on the right-hand side of the screen. There is a vector layer, meaning Krita will recognize the `.svg` elements on this layer. No vector data will be lost.

The vector shapes on the layer are selectable and editable, and you can move and resize them, almost as in Inkscape. Krita is not a vector editor program by default, but it can handle the `.svg` elements with ease. This is why it works so well with Inkscape illustrations:

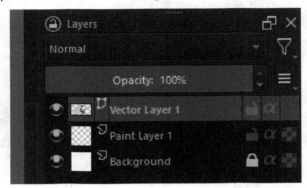

Figure 7.3 – The pasted SVG shapes are based on a vector layer in Krita

Also, notice how after pasting it into Krita, the vector image seems smaller. It is not fitting the background, even if you set it to the same size in Inkscape as the image size in Krita. The reason behind this is that Inkscape is working in a default 96 dpi resolution, while this document is 300 dpi. That is why the drawing seems smaller, but it is easy to fix that.

Don't forget that this is a vector illustration on a dedicated vector layer, and that comes with all the benefits of working with vector elements. You can edit, transform, and resize the illustration to any size without loss.

Use the **Select Shapes** tool and click on the illustration. All the shapes will be selected, because these elements are in still one group in Krita, as they were in Inkscape.

Of course, Krita has different hotkeys than Inkscape. So, to resize the image to fit the background while keeping the proportions, you need to hold the *Shift* key instead of the *Ctrl* key as in Inkscape:

Figure 7.4 – This is how the .svg file looks when you select it in Krita

Feel free to resize and reposition the vector image so that it matches the document size before moving on to the next step.

Adding shadows and highlights to your illustration

The vector part of the illustration is scaled and in position. It is time to add the visuals that made us use Krita in the first place!

The vector elements will be the basis of this illustration. They will provide the colors and the original shapes and structures, and we will not change them for now. What you will do is paint shadows and highlights on a layer over the vector illustration.

To be able to do that, first, locate **Paint Layer** on the **Layer** tab (you can see it in *Figure 7.3* as well) and move it above **Vector Layer**. Click and hold the left mouse button on the layer in the list and simply drag it above **Vector Layer**. Layers work the same way in all design programs: whatever you draw and paint on the layer above will be visible in the image, covering the elements on the underlying vector layer.

Before painting the shadows, you will have to select the part of the vector illustration you will add the shading to. This selection is needed so that the freehand shadow or highlight will stay within the bounds of the original shape.

Let's start with the face of the lady. Click the **Contiguous Selection** tool (or *Magic Wand*) to select the shape of the face. We could select the individual shapes as well, but this tool selects everything that is connected and has similar colors. So, click on the face in the **Vector Layer** area and select it.

Now, switch back to the **Paint Layer** area and click on the **Freehand Brush** tool on the toolbar. The brush presets are located in the lower right of the screen. Krita has a wide variety of texture brushes that can imitate realistic art tools such as pencils or watercolor brushes, as well as specialized stamps to add grunge effects or falling leaves to your image. For this task, I suggest you use the soft airbrush, a simple brush for creating subtle textures.

After selecting the airsoft brush, at the top bar, set **Opacity** to **20%**, and the brush size to about **150 pixels**. You don't have to use these exact numbers – we just need to keep the opacity low and the brush bigger, to get a subtle and smooth effect at the end.

Since you want to add a shadow here, you will need a darker color. To pick the exact color of the vector shape, hold the *Ctrl* key and left-click on the screen for an instant color picker.

This will set the color for all the tools now in Krita. And if you look at the top-right corner of the screen, you will notice the same color wheel we are used to in Inkscape. Move the marker in the color triangle and select a darker shade of the color you just picked from the vector shape.

Now to the actual coloring: draw the shadow you want and don't worry about running out of line! You will not paint over the other shapes accidentally. This is why we have the selection in place. Have fun while painting the shadow; just apply some darker skin tone on the left-hand side of the character's face. Check out *Figure 7.5* for my results:

Figure 7.5 – Simple shading added to the face of the character

When you feel that the shading here is finished, you need to deselect the shape with *Shift + Ctrl + A*. It is an important step to see your image without the dotted line of the selection. If you deem it not ready, press *Ctrl + Z* to reselect the shape again, and paint a bit more.

This looks smoother than the shading we did back in *Chapter 5, Edit a Photo and Create a Hero Image in Inkscape*, and it is also freehand shading. This is why we are using Krita now.

There will be a lot of switching back and forth between the two layers from now. You will select the shape on the vector layer, then apply a shadow to the **Paint Layer** area. Let's shade another part of the image: the hair.

Jump to the **Vector Layer** area and select the shape of the hair with the **Magic Wand** tool. Notice how the hair is selected as one shape, though it is built up from two separate vector shapes. The cause of this is the mechanics of the **Magic Wand** tool selecting all the same – or similar – colored pixels in place, as explained previously.

Now, go back to the **Paint Layer** area and click on the **Brush** tool. Of course, Krita will remember your settings, so the **Airbrush** tool will be set up as you left it.

However, you will need another color to create a shadow on this bright red surface. Again, hold down *Ctrl* and click on the hair to pick its red color, then modify this color with the color wheel to get a slightly darker shade of it.

Now, apply a smooth shade to the hair as well. Of course, click more than once, and use this soft and almost transparent brush like a real airbrush. Build the tone up with more and more touches. Add more layers of color behind the head at the bottom of the hair and leave the top almost intact.

If you feel you did it too much, then hold down the *Ctrl* key and click the hair again to get back the original color and paint over it with that. If you are following the shape, you can add a subtle shadow effect to your image, as shown in *Figure 7.6*:

Figure 7.6 – The hair of the character is also shaded softly

The next part should be the torso of the character. As you may recall, we were playing with negative space there, while drawing it in Inkscape. This means that the sleeve of her T-shirt is not really there; we just make it appear like it is by cutting the shape of the arm in a straight line.

You will do something similar now in Krita but emphasize the sleeve a bit with the help of shading:

1. First, select the torso with the **Magic Wand** tool (named **Contiguous** selection on the toolbar).

    This will select the whole blue area of the torso. If you were to paint the shadows now, it would paint over the whole shirt, including the sleeves. Since you don't need that part to be painted now, you will subtract that area from the selection.

2. Choose the **Polygonal** selection tool (just above the **Magic Wand** tool in the toolbar) and draw a simple shape above the selection while holding *Alt*. To subtract it from the original selection, you will have to hold down the *Alt* key when you close the shape. After finishing the shape, you will have to have a selection with a stripe unselected.

3. Hold down the *Ctrl* key and click the blue surface to pick its color and select a darker shade of that using the color wheel. Now, if you start painting the shadows on the shirt, the arm will not be painted, only the parts above it and under it.

This will create a sense of depth in the illustration. After finishing the shading, you will end up with a shadow similar to the shape shown in *Figure 7.7*:

Figure 7.7 – Shading the torso after using the Polygonal selection tool

Lighting can go very far; it can go into endless details. The shapes could work like this, but you might add a lighter color as well, to create highlights and simple lighting. You could paint the lights on the same layer as the shadows, as well as add a different layer for the lights only.

Repeating the process for all of the elements

So far, you have added simple shading to three surfaces in Krita. Since this book is not about Krita, I will stop here, and let you finish this illustration on your own. You will need to repeat the same techniques all over. The steps will be easy to follow; you will have to repeat these for all of the elements:

1. Select the shape on the vector layer using the **Magic Wand** tool.

2. Pick the color of the shape and mix a darker/ brighter version and paint the shadow/highlight on the paint layer above the vector layer.

3. Deselect by pressing *Shift + Ctrl + A* and repeat this process for the next shape in the illustration.

And that is it – repeat these steps to practice the process of shading a flat illustration in Krita. If you keep on adding shadows and lights to the elements, you will end up with an illustration like the one shown in *Figure 7.8*:

Figure 7.8 – The final illustration with shadows and highlights

Continue with this process to create the results you are looking for.

Finishing the project in Krita

To finish this project, just keep on adding shadows and highlights. Of course, you can go much further than what we did in this example. You could add textures, use filters, and paint additional details on the illustration using the same method.

When you are done, save your work as a KRA file. That will include the editable SVG you used as a base, plus the paint layer(s) and settings you created. If you want to print this work, export it in any high-resolution bitmap format: Tiff, JPG, or PNG.

Krita is a great app, and easy to combine with Inkscape this way. Of course, instead of Krita, you might use any other drawing program that supports layers and vector elements. You could have almost the same result and similar process using Gimp or Adobe Photoshop.

Hopefully, this small project inspired you to combine Inkscape with raster painting programs and create quality illustrations with more details!

Inkscape and Scribus – desktop publishing with crisp vector elements

After painting over vector illustrations with Krita, the second project in this chapter will be less about art, and more about presenting content. To be more precise, this project is about paring Inkscape with the open-source desktop publishing software called Scribus.

It is a well-equipped program that allows users to create different text layouts from single-page presentations to whole books with hundreds of pages of text and illustrations.

Crisp vector images are perfect for printing – and that is why Inkscape fits so well with Scribus. The developers of Scribus know this too, which is why they added great SVG support to their program.

One of the few shortcomings of Inkscape is that you cannot export print-ready files in CMYK color format from your artwork. During this project, you will also learn how Scribus offers an easy solution for this, completing the toolset of Inkscape.

Also, desktop publishing and editing flyers and booklets or longer books is a pain in Inkscape. Learning the ropes of Scribus during this short project might be useful for you later.

> **What will you create in this project?**
> Export the complex vector illustration from Inkscape and use it as a vector image in a Scribus document to create a simple flyer for CloudUsers.

In this project, Inkscape will provide the visual elements, while Scribus will handle text, and anything related to the structure and functionality of the document.

Setting up the document in Scribus

So far in this book, everything started with Inkscape. This could be the same with this project, as you will prepare an illustration in Inkscape and use it in Scribus. But since the result of this project has to be a print-ready flyer, let's start with Scribus and set up your document:

1. When you create a new document in Scribus, a window will pop up with several settings.

2. Since you will create a simple flyer now, pick **Single Page**.

3. Set **Orientation** to **Portrait** and set **Width** to 4 inches or **99 mm**, and the **Height** to 9 inches or **210 mm**.

    These are the standard sizes for rack cards. A rack card is a small flyer you might see on hotel countertops or handed out at events or a tall flyer with a small amount of information on them.

    For printing purposes, you should also set up the bleeds now. This will set the trimmed size of the document; everything over the bleed area will be cut off after printing.

4. Locate the **Bleeds** tab and set it to a standard 0.125 inches (**3 mm**) for all sides. If you click the link icon next to the measurements, it is enough to set **Bleeds** up once. Leave the margins at their default values for now:

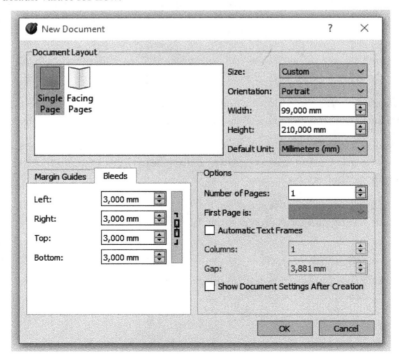

Figure 7.9 – The settings for the new document in Scribus

When you are finished, press **OK**. You will see a blank document with a red and blue frame inside it. The blue one signals the margins, while the red one is for the bleeds.

> **Tip**
>
> You can set the sizes and other settings at the start or later from **File | Document Setup**. Scribus is a desktop publishing software after all, which means all the settings can be changed anytime without quality loss, just like in Inkscape.

Adding a text frame to the flyer

The flyer will have some neat vector graphics of course, but before that, you have to set up the text frame. In a nutshell, in Scribus, text is added via **Text Frames**, not as individual lines. These boxes hold paragraphs of text, and each paragraph can have different styles applied to it. Styles are font sizes, weights, colors, and countless other settings that are reusable and persistent throughout the document. Follow these steps:

1. To add a **Text frame**, go to **Insert | Text frame** in the top menu, click the *T* icon in the toolbar, or press *T* on your keyboard.

2. Then, draw your text box around the middle of the document. This might sound vague for now, but you can resize and reposition it later if needed.

3. Right-click on the text frame to fill it with a bit of sample text. Choose **Sample text** from the popup menu and select **Standard Lorem ipsum**. Two paragraphs will be enough for now. Press **OK**; the frame will be filled.

4. Double-click the text frame and type in some header text right above the sample text. You might use the same sentence we used while designing the website in *Chapter 6, Flexible Website Layout Design for Desktop and Mobile with Inkscape*.

> **To our users, from the cloud**
>
> **Scribus** is a publishing tool. It is all about formatting text and has many more typography settings than Inkscape. As mentioned previously, styles are very useful while editing long documents, but are a great way to set up the text here too.

5. Press *F3* or choose **Edit-Styles** from the top menu. This will open the **Style Management** dialog. Here, you will see a default paragraph style already set up. This will seem overwhelming at first, but you will only change a few things:

    I.   First, in the **Properties** tab, set **Alignment** to **Justified**.

    II. Then, in the **Character Style** tab, set **Font** to **Work sans** and **Font size** to **12** points. That is all for this text style!

6. Now, create a new style for the header text. Click on the **New** button and select **Paragraph Style**. Name the new style Header so that you can find it later. Then, as before, set the paragraph and character styles:

    I. On the **Properties** tab, set the alignment to left, and the line spacing to **38pt**.

    II. On the **Character style** tab, choose **Montserrat** and set **Font Weight** to **Bold** and **Font size** to **46pt**.

7. When you are finished, press **Done** to save the style settings:

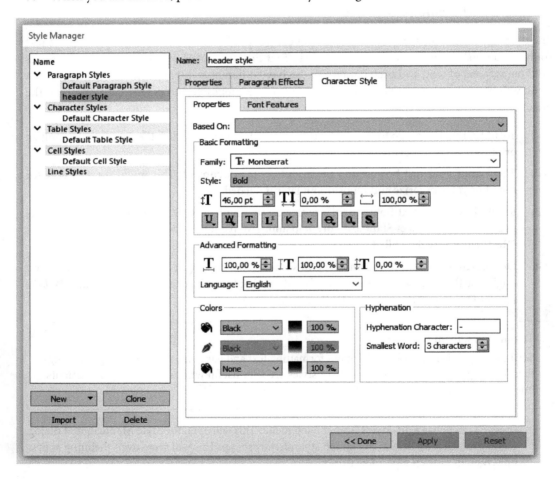

Figure 7.10 – The Style management dialog with the Header style settings in Scribus

There will be no changes visible in the text frame yet because the styles have to be applied to the text. Let's learn how to do this.

8. Select the text frame with the sample text and go to **Edit | Edit text** or simply press *Ctrl + T*. This will bring up the **Story Editor** dialog window. As you can see, the different paragraphs are separated by lines on the left. No styles have been applied to the paragraphs yet, but you will change that now.

9. Click on the **No styles** prompts on the left and choose the **Header** styles for the header line from the dropdown list. Then, do the same for the other paragraphs, using the **Default paragraph** style this time.

10. When you are done, click the checkmark icon in the **Story Editor** area to apply the changes and close the editor window. Your results must be like those shown in *Figure 7.11*!

Figure 7.11 – Text frame with sample text and a header in Scribus

This is where we will leave Scribus for now, so save the document. You will come back to it after getting the visuals from Inkscape.

Preparing the visual elements in Inkscape

The Scribus document is set up, so it's time to get Inkscape into this workflow. As with the other two projects in this chapter, we are not focusing on creating new work in Inkscape, but rather building what you built during the previous chapters.

For this particular project, you need to get the CloudUsers logo, and the complex illustration you created with the characters in the cloud.

Open the documents with those elements inside. If you have not created these elements so far, you can find the relevant projects in *Chapter 2, Design a Clever Tech Logo with Inkscape*, for the logo, and *Chapter 4, Create Detailed Illustrations with Inkscape*, for the illustration. Of course, you are free to use other SVG illustrations as well for practice purposes.

Group both sets of elements. The logo and the illustration can stay in separate files, of course, you are just grouping the parts to make sure nothing is lost while moving the illustration to Scribus.

There are two ways to import SVG objects into Scribus. You can save and import the illustration into a new SVG file or copy and paste the elements from the original file at hand. The second method does not require much preparation of course, so let's learn about the first one a bit:

1. Copy the illustration and paste it into a new blank document in Inkscape.
2. Set **Page size** so that it fits the drawing. The easiest way to do this is by selecting the drawing and pressing *Shift + Ctrl + R*.

After this, save the file in plain SVG format. This strips the file from its Inkscape metadata and helps Scribus handle it correctly (this is the best route for other programs as well).

The same happens when you copy the elements into Scribus; they appear as plain SVG. But there might be more errors or unwanted elements, things missing, and so on. But since Scribus 1.5, the SVG handling is almost flawless.

Importing and working with vector elements in Scribus

Previously, we mentioned two ways to import vector shapes into Scribus. It is up to you which method you use.

If you chose to save separate plain SVG files to import them into Scribus, then you can go to **File | Import | Get vector file**, select the file, and click on the page where you want the vector illustration to be imported. This method is might be cleaner since the files are more organized.

If you chose to simply copy and paste your vector elements, then just select the groups you need in Inkscape, press **Copy**, and then **Paste** them onto the page in Scribus. This method is more convenient, but it might be harder to control the SVG file this way.

Whichever method you use, the results will be the same: the SVG vector elements will be on the page now.

> **Tip**
>
> Scribus works well with the .svg format, since vectors are great for printing, and the developers implemented SVG support. But there might be some limitations you might come across. It is better to be proactive and work them out in Inkscape, rather than in Scribus with a limited vector toolset. Always check your vectors after importing them into Scribus, and if there is any difference, see if you can solve it. It might happen, for example, that objects lose opacity, or you miss an object while copying and pasting your illustration.
>
> Also, keep in mind that effects such as noise and blur will not be copied over since they are not standard SVG vector elements but raster effects! Scribus will not work with them and drop them from the SVG!
>
> If you have a design with filter effects, export it as a .png file and import that as an image. Of course, in that case, you will lose vector edibility, and you will have to export them in a high resolution for printing (a step you do not need for vector shapes).

When you insert a group of SVG objects into Scribus, they will behave as basic shapes such as polygons, circles, and horizontal separators that you can draw in Scribus to enhance your page. You can ungroup them, scale them, set their transparency and color, and individually move the elements around the page.

It is more complicated to draw complex vector illustrations in Scribus (this is where Inkscape comes into the picture) but you can modify them to an extent with ease. And this is what you will do now:

1. Move the cloud illustration to the bottom of the page. This will make the laptop sit on the *ground*, and the big blue cloud will give the flyer a bit of weight.

2. Now, ungroup the illustration by right-clicking and selecting **Ungroup** or pressing *Shift + Ctrl + G* just like you would in Inkscape.

3. Ungroup it once or twice until the girl on the cloud is separated into one group. Move the *cloud girl* aside from the page and ungroup it again.

4. Delete the down arrow since this will act like a separate visual element now, and the arrow does not help the composition.

5. Select the elements of the *cloud girl* and group them again by selecting **Group** from the right-click menu or pressing *Ctrl + G* (again just like in Inkscape).

6. Select the rest of the big illustration and group it back again. Now, you have two groups: the girl on the cloud and the rest of the people on the giant laptop.

7. Position the *cloud girl* group upwards and to the right so that it is almost filling the space on the right of the title text. You can partially cover some letters (as shown in *Figure 7.12*) to create an interesting composition for the flyer. Scale it by holding *Ctrl* as needed to keep the proportions:

Figure 7.12 – Part of the illustration over the header text in Scribus

8. Now, let's get back to the rest of the illustration: move it to the bottom of the page and scale it that so most of it is on the page, the laptop is in the middle, inside the margins, and the people are visible.

It is OK if the big cloud is outside of the margin and the bleed guides! It will be cut and make the flyer even more interesting since the illustration runs over the edges.

Just like Inkscape, Scribus also allows you to level objects, meaning you can easily send this illustration behind the main text. Let's learn how to do this:

1. Select the text and choose **Level** from the right-click menu. Send the illustration to the back so that it appears partially behind the text. (See the final design in *Figure 7.13* for guidance!)

2. To check how the final position and size of the illustration will look on the paper, press *Ctrl + Alt + P* for **Preview mode**. This mode will show how the bleed works; the image is virtually trimmed at the edge of the bleeding, where the paper will be trimmed after printing. Press *Ctrl +Alt + P* again to get back to **Edit mode**.

Now that the illustrations are in place, it is time to add the logo to the flyer:

1. Import the CloudUsers logo – either importing it as a `.svg` vector file or just pasting it straight from Inkscape. Your task is easy now – position and scale the logo on the top of the flyer.

2. Match its size and position to the blue margin guides, and stay inside them this time. We don't want part of the logo being trimmed or touching the edge of the page. Refer to *Figure 7.13* for the final logo position!

    The visual elements of the logo are all in place; now, the only thing that seems off is the color of the text.

    You will have a bit of work with the color settings soon, but before that, you need to fix the color of the header and the main text so that it matches the overall brand colors. Since you set up styles to design the text style of the flyer, you will have an easy job.

3. Press *F4* to pull up the **Style manager** area. Head to the **Character Styles** tab of the **Header** style.

4. Locate the **Color** settings at the bottom of the window and click on them.

    You will notice that, among the few default colors, there is a list of familiar colors with names starting with **FromSVG#**. These are the colors of the imported SVG elements that are now part of this Scribus file. These colors are now saved and can easily be assigned to other elements in this document.

5. Pick a darker blue for the **Header** text (#00aad4), then click **Apply** to see the changes on the document.

6. Without closing the **Style** manager, repeat the same for the **Default** paragraph style. Set a darker text color (#044b5d) in the **Color** dropdown in the **Character Style** tab. The final result is shown in *Figure 7.13*:

Figure 7.13 – This is how the final version of the flyer will look after being printed and trimmed

With the colorful characters and the text in place, this looks like a freshly illustrated card! The vivid colors are important for the brand, and to keep those colors after printing, you need to set up the final version of the file properly!

Saving and exporting your finished flyer for printing

If you want to hand this flyer to a printing company, they will ask for a file using CMYK color format. Inkscape does not include CMYK color information in the `.svg` files by default – primarily because SVG was meant as a digital format for simple online graphics.

But luckily, Scribus can handle different color profiles and print-ready file formats. Any SVG elements that are used in Scribus can have CMYK color information assigned to them and get ready to be printed professionally. Let's get started:

1. First, open the **Document Setup** area and locate **Color Management**. Check the **Activate Color Management** checkbox and press **OK**.

2. With **Color Management** enabled, you have to set every RGB color to a CMYK equivalent.

3. Go to **Edit** and pick **Colors and Fills**. There, you will see a list of the **FromSVG#** named colors that you saw earlier while coloring the text. Next to the color names, there is a little RGB or CMYK icon, which indicates the color version. Double-click the color and set it to **CMYK** and press **OK**.

4. Repeat this to every color in the document, then press **OK** and save:

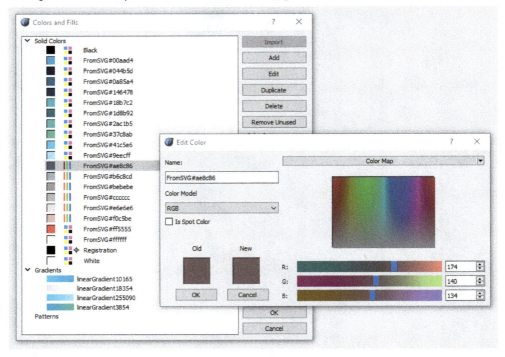

Figure 7.14 – Setting up CMYK colors from the SVG colors in Scribus

Now, you are ready to export your flyer as a CMYK PDF!

5. Go to **File** | **Export** and select **Save as PDF**. Set **Profile** to **PDF/X-4** and press **OK**. There are plenty of PDF settings in the dialog window that pops up: You can decide to embed fonts, add crop marks and color marks for the printer, or password-protect your PDF.

What is important in our case is to check the **Pre-press** tab and make sure to use the bleed from the document, as well as check if the exported PDF will use the CMYK color profile you defined earlier.

If all of these are set correctly, just press **Save**; your file will be saved as a print-ready vector PDF!

Of course, this was a simple task, but hopefully, this example project was enough to show you how Scribus and Inkscape can work together! You can upscale the same methods learned in this project to create longer and more complex documents by combining the SVG power of Inkscape with Scribus' publishing capacities.

There are many uses outside of the scope of this book, but if you are interested in desktop publishing, I suggest learning more about Scribus and taking advantage of its printing and editing features!

Inkscape and Blender – turning your 2D work into 3D!

The third project in this chapter is something different. This time, we will not design anything flat but use Inkscape in a 3D workflow. Inkscape is not a stranger in the land of 3D; many designers use Inkscape to create textures, decals, or technical plans to work with Blender. Blender is an open-source 3D modeling and animation software.

If you use Blender already, you might like to use Inkscape to create easy-to-use elements to work with. If you are new to Blender but have followed the projects in this book so far, you can still learn something and broaden your design knowledge with this short project.

> **What will you create in this project?**
> Export the vector logo of CloudUsers from Inkscape, then import it into Blender and create a 3D object out of it.

Preparing the SVG in Inkscape

Again, we will start this process with Inkscape. Open the file that you created during *Chapter 2, Design a Clever Tech Logo with Inkscape*, that contains the CloudUsers logo. You will export this logo from Inkscape and import it into Blender. If you do not have the logo yet, you can create it by following the steps in the aforementioned chapter. Alternatively, you can use any other simple SVG illustration or logo you have created.

After we created the logo, we created different versions of it too, as we usually do before handing over the logo to the client. Among those versions, there was one with the text under the emblematic part of the logo, as shown in *Figure 7.15*. Copy that version and paste it into a new document:

Figure 7.15 – The CloudUsers logo you will use in Blender

In the new document, select the logo with the text, and go to **Path** | **Union** to merge it all into one object. This will help later on in Blender.

An SVG file imported into Blender is projected on the horizontal plane. Although it maintains its color, Blender cannot read gradients or transparency from the SVG file. And of course, it loses all raster effects, such as blur or filters.

To prepare for this, whenever you intend to use an Inkscape vector in Blender, make sure that it has no irrelevant color data. Select the logo and set it to the blue of the CloudUser brand as we defined earlier. The code for that is #18b7c2.

After the colors have been set, resize the page to the drawing by selecting the logo and pressing *Shift + Ctrl + R*.

When you are done, save the SVG file as a plain SVG in Inkscape. This will remove the Inkscape metadata from the file and keep the information that is valuable to Blender.

Using Blender to create a 3D version of the logo

SVG files are the only vector files that you can import into Blender. They are easy to transform and luckily, easy to create with Inkscape:

1. To import your logo file, go to **File** | **Import** and select **Scalable Vector Graphics (SVG)**.

2. Navigate to your file location and click **OK**. If you do not see any changes in Blender, it might be that your SVG file is simply very small.

3. Select all the parts in the middle of the screen, then press *S* and scale your SVG up a bit (or a lot) if needed. Since you merged all the parts into one path in Inkscape, you do not have to work on all the objects individually.

It will only show the one path to edit:

Figure 7.16 – The logo SVG imported into Blender

The vector data is all here, and to be able to edit the SVG in Blender, you need to convert the path into a mesh.

4.   Right-click the logo and select **Convert To | Mesh**.

3D meshes are a system of vertices, edges, and faces that build up any 3D object. Any of these elements of a mesh can be selected, modified, moved, or deleted. This is how 3D objects are shaped in modeling programs.

With the SVG path converted into a mesh, the nodes of the path turn into vertices on a 2D plane. There will be more vertices than nodes to describe the object with a mesh. More vertices means more surfaces, and more possibilities for errors later. So, as a next step, you need to clean up the mesh a bit.

Usually, cleaning up and merging surfaces is enough to prepare the 3D object.

5.   In **Edit mode**, at the top-left corner of the Blender screen, you can choose between selection methods. Choose the third icon for selecting surfaces. Select the surfaces in one part of the logo and press *F*.

This will merge neighboring surfaces and, as shown in *Figure 7.17*, clean up the surface:

Figure7.17 – Cleaning up the mesh

Most parts of the mesh can be merged into a single surface. The only tricky part is the letters, such as *R* and *O*; they need to be on two to three different surfaces to keep that hole in the middle.

After the mesh is clean without lone vertices and broken surfaces, it is time to turn this 2D logo into 3D. There are two different methods to do this in Blender. Both work well, but I think they produce different outcomes:

- The first method is to extrude the SVG shape along the vertical axis. To do this, select all the surfaces on the logo, and press the **Extrude Region** icon in the toolbar on the left.

  Then, extrude the surfaces vertically in one move. That's it – this is a simple move, and you have a 3D object ready.

- The second method is to add a modifier to the mesh. Blender has a lot of these by default, and they can be applied from the **Property Editor** window:

  I.   Select the logo mesh, then go to the **Modifier Properties** tab of the **Properties** editor (the wrench icon, as shown in *Figure 7.18*).

  II. Then, select **Add modifier** and select **Solidify** from the list.

  This will give your 2D mesh thickness. Maybe the first method is faster, but the **Solidify** modifier can be edited later on. You can manually set the thickness, add various other properties, and apply other modifiers as well!

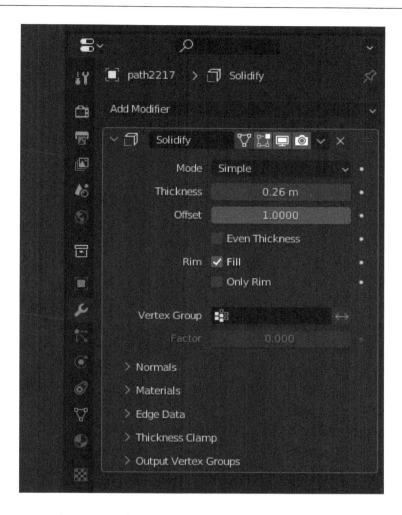

Figure 7.18 – The modifiers panel on the right of the screen

You may choose either of the two methods; the results are similar. Once you've done this, your 3D logo will be ready to be animated, rendered, or prepared for 3D printing! See the results in *Figure 7.19*!

Figure7.19 – The final 3D object

Hopefully, this glimpse into Blender added yet another point to the list of the ways SVG files – and Inkscape – can help you on your creative journey.

Summary

In this chapter, you got a taste of three different fields where Inkscape and SVG files can be used. The goal was to show you the possibilities without going into too many details for each program.

First, in an artistic project, you learned how to paint shadows over Inkscape vector illustrations in Krita. This included using vector layers and raster brushes to create soft visual effects. Then, you had a run with Scribus and created a simple flyer for CloudUsers.

The goal was to learn how flexible Scribus is with vector elements, how it works with text and layout elements, and how to export CMYK files. And in the last short section, we looked into Blender for a change, learning that Inkscape can provide a great start for 3D modeling as well.

This chapter aimed to open up new ways you might think about Inkscape. It is created for vector image editing but can be a starting stone for creative projects from different fields.

In the next chapter, we will learn all the small tricks and tips you need to know to use Inkscape like a pro every day!

8

Pro Tips and
Tricks for Inkscapers

This chapter is a collection of tips and tricks. It is based on working every day on real-life design projects with Inkscape. These are tricks that did not fit into any of the projects we have presented but are crucial if you want to become an Inkscape power user. These methods can help you become more effective and resilient and hopefully answer questions that you still have after reading this book.

Also, these methods are mentioned as alternatives in other parts of this book, so if you are here looking for those tricks, please read on!

In this chapter, we're going to cover the following main topics:

- Practicing with the basic tools
- Praise for path effects
- Inkscape as an XML editor
- Overcoming the CMYK color barrier in Inkscape
- Working faster with custom templates
- What to do if Inkscape crashes

Practicing the basic tools but looking further

It is almost a cliché to state this, but the old saying is true: *practice makes perfect*. This is the first pro tip for you as a fellow Inkscape user.

You first have to understand the basic tools of Inkscape and use them regularly before getting into clones, extensions, filters, and other tricks. You can only utilize these properly if you know your tools. Practice with the Inkscape tools and learn the fundamentals of vector graphics. Learn the basics before overwhelming yourself with the rest.

So, what are those basic tools and methods that are worth knowing and practicing?

Here is a very short list of what you should know when using Inkscape:

- Creating and transforming simple objects
- The usage of colors, gradients, and blur
- Working with paths and using Boolean operations
- Layers and groups for organizing
- Understanding clipping and masking

These were the basics, and after that, you can move on to more complex tools and methods. If there is something you are not familiar with from the preceding list, take your time, get to know that part, and practice. As you know, this book is about specific projects and not introducing each tool to the user one by one.

We started with simple projects and worked our way toward the more complex ones. The goal is not to regulate the way you learn. The goal is to help you better understand vector graphics.

Looking beyond the basics

Until now, all the previous chapters shared examples of real-life projects. These projects were all about building the basic tools of Inkscape into a logical workflow to create a logo, an icon set, or a website design.

Knowing the basics is important, but you still need to look further. Be open to learning new tricks and looking into new solutions to problems. Only this will make your work of higher quality and more efficient.

In this chapter, there will be no step-by-step projects to follow, just tips and very short examples of using some of the less commonly used tools in Inkscape and using Inkscape in uncommon ways.

Praise for path effects

After practicing the basics, you will work faster and will be able to create original work in Inkscape. But a lot of times you need to be faster, or the work is more complex, and you start to look for semi-automatic solutions. One of the best tools that can help you in these situations is **Live Path Effects** (**LPEs**).

LPEs are effects that you can apply to any selected path. Whenever the original path is modified, the effects change with the path in real time – hence the name.

LPEs can make your work faster and easier, and they are a lot of fun to use. My suggestion is to play and experiment with LPEs and find the ones that work for you. You will find different effects to be useful for different projects.

Using a simple path effect – Bend

To bend a shape in Inkscape manually, you need to change the path node by node with the **Node** tool. It is not an easy task; you have to select and move the nodes around and have a good idea about what your end goal is.

This is where the Bend effect can come into the picture, helping you to stretch and bend shapes with ease. It is a great path effect to try first since it is very easy to use and understand.

Let's give it a try now:

1. To start, you will need a path. It can be a simple circle turned into a path, a random shape drawn with the **Pencil** tool, or something you drew earlier following the projects of this book. Draw anything, but if you want to follow my example, use the cloud shape we created in *Chapter 3, Modular Icon Set Design with the Power of Vector*, while designing the cloud icons.

2. Now click on **Path | Path Effects** in the top menu. This will open the **Path Effects** tab on the sidebar.

3. Select the path you want to apply the effect to, and click the + button at the bottom of the **Path Effects** tab. The **Live Path Effect Selector** window will pop up, showing an array of the available path effects to choose from. This first time, choose **Bend**.

4. As you can see in the **Path Effects** tab, this applied the Bend effect to the path.

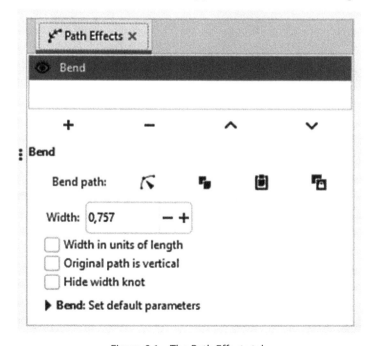

Figure 8.1 – The Path Effects tab

5. On the path itself, seemingly nothing changed. But if you select the **Edit on-canvas** icon, which is the first icon option in the **Bend** path line, you will be able to bend the whole path by curving the new *spine* path with the **Node** editor tool.

6. Create different curves to see how the original path reacts and changes. You can even duplicate the path and change the amount of bend creating different objects from the same path.

This effect can make vector graphics even more versatile. Draw a character and easily bend its leg, distort many leaves of a tree to seem more unique, or create your own crazy text effects.

The most important thing is: LPEs are non-destructive. Using them for editing does not modify the path permanently. You can remove them with the minus icon as easily as you added them to the path. The original path will always be intact.

Take the clouds in *Figure 8.2* for example; remove the path effect, and all of the paths will pop back to the original shape.

Figure 8.2 – The original cloud path and more versions of it with the Bend effect

However, if you want to make your changes permanent without applying LPE, you can do that too.

Just select the shape and turn it into a path by selecting **Path | Object to Path**, or pressing *Shift + Ctrl + C*. What this does is apply the changes you created to the shape and turn it into a new path. The original path will be lost, but the new one will be there without an active path effect.

What do you gain by using this method? A simple path might use fewer resources without active path effects applied, thus making Inkscape more stable when working with your file. Also, the changes will be there even if the file is opened in another vector editor, and they will look the same as intended.

> **An important tip about LPEs**
>
> Path effects are great, but when applied to complex shapes with hundreds of nodes (or groups with several objects) they can use a lot of resources. Save often to avoid losing your work if Inkscape crashes. If that happens, check out the section in this chapter titled *What to do if Inkscape crashes*.

Although LPEs are *path effects*, you can apply them to groups of objects as well! This means you can bend, distort, or modify a group of objects as one with a path effect, even if the elements in the group are not all paths! You can try this now with the Bend effect, or with the Perspective effect next.

The Perspective effect

LPEs are very effective and can really help your work. Our second example, the **Perspective** path effect, can also spare a lot of headaches. It is much easier to draw something in frontal view and then modify it into perspective than drawing it straight on without a reference.

This path effect is simply great for drawing houses and cityscape backgrounds or applying illustrations to surfaces that are not perfectly facing the view.

Since LPEs work on groups, too, let's try the Perspective path effect on a group first. You might create a group of a few objects to try now or use anything you have created with this book so far: an icon, the logo, or any illustration.

> **Note**
>
> If you are through *Chapter 4, Create Detailed Illustrations with Inkscape*, and looking for an alternative solution to fit the image on the laptop screen, here it is!

If you want to try it on something real, try it with the complex illustration from *Chapter 4, Create Detailed Illustrations with Inkscape*! This is the example we will use now.

1. To start, open the complex illustration you created earlier, and select the group with the lady on the screen. We used a different method in *Chapter 4, Create Detailed Illustrations with Inkscape*, but the Perspective effect works even better!

2. Select the group with the flat image of the lady and add the path effect by selecting **Path | Path Effect** in the top menu.

3. Click the + icon at the bottom of the **Path Effect** tab.

4. This time choose the **Perspective/Envelope** effect from the **Live Path Effect Selector** popup. Seemingly nothing happens, but on the **Path Effect** tab, you can see that the Perspective effect was added to the effect list.

5. Still selecting the group, switch to the **Node** editor tool, and notice how a rectangle with four nodes at the corners appears around the group. See *Figure 8.3* for reference!

6. Move the four nodes around to create the Perspective effect. In this case, we have it relatively easy. All you need to do is match the four corner nodes to the corners of the screen of the laptop.

Figure 8.3 – The Perspective path effect applied to the group to create a screen

That's it, it is easy! Apply the Perspective path effect and set the corners. If you need to be very precise in a future project, you might set the X and Y coordinates of the four handles that define the Perspective effect.

There are many more path effects

The Bend effect and the **Perspective/Envelope** effect are just two of the useful path effects that you might utilize in Inkscape. There are more LPEs that we will not mention now since this book is not about trying all the tools. You have to do that yourself; use them and experiment with all the different effects.

There are many more, but here are a few I use regularly:

- **Pattern along path**: With this, you can use single shapes to create a repeated pattern along any path. Think further than dotted lines; you can create linear patterns of anything to enhance your vector illustrations. Use this effect with your own paths pasted from the clipboard, and create footsteps, chains, zips, and so on.

- **Power stroke**: This effect is there to help you create really energetic strokes that are great for illustrations and line artwork. You can manipulate the shape and thickness of the strokes after they are drawn, allowing you to use really dynamic strokes, almost like using a brush.

- **Corners**: This effect lets you add rounded corners to rectangles and polygons, and it is a great help – I use it to round the pointy corners of stars or set different corners for buttons or other UI elements.

- **Envelope deformation and Lattice deformation**: These effects are quite similar to each other. In a way, they combine the Perspective and the Bend effects, allowing you to modify a path or a group even more fluidly. They are great for creating waves or patterns on clothes or flags.

- **Tiling**: This is the latest great addition to the LPE library. Introduced in Inkscape 1.2, the Tiling LPE can help you create tiled patterns from a single selected path. It can be rotated, scaled, or mirrored, you can set the gap between the objects and, of course, randomize your tile pattern if you want to. This is a great automation when you need to create intricately tiled backgrounds for pattern design or for illustration purposes.

In summary, LPEs are fun and useful. Check them out and try using them in your Inkscape projects. Find your favorites and use those in your daily workflow.

My list of favorites is always changing according to the project I am currently working on. My favorite LPE is always the one that can help me be creative in the fastest and most simple way possible.

In this section, you learned about a few useful LPEs and how to use them. In the next section, we will focus on the XML-based tricks you can use to work even faster in Inkscape.

Using Inkscape as an XML editor

This book is all about design projects and creating vector graphics in Inkscape, but sometimes you can do a bit more and use the SVG format to your advantage! SVG is an open standard format, which means you can look into it and, with a bit of work, understand it.

It uses the XML format to describe vector graphics, so you can modify SVG files even with a simple text editor. Don't be alarmed; the goal is not to build SVG graphics from scratch via XML! That would be counterproductive; that is still the job of Inkscape.

In this part of the chapter, I simply want to share a few tricks where the XML editing capabilities of Inkscape can be at your service.

Searching for, selecting, and changing multiple elements at once

Imagine working with a file with hundreds of different objects and shapes, such as small leaves on a tree illustration or multiple buttons and icons in a complex UI design. And then, you need to change the color of every other object in the file. Not all of them, but enough of them that selecting them by hand would be a real pain.

This is where the simple **Find/Replace** tab comes to the rescue! It may sound trivial, but this is an overlooked part of Inkscape with great potential.

Let's give it a try now. The icons you created during *Chapter 3, Modular Icon Set Design with the Power of Vector*, will be great test subjects. Although there are only nine of them, you will be able to use the same method later for more objects.

For the sake of practice, let's say that you need to change the turquoise color of the icons to orange. You could select all the turquoise parts manually one by one, then enter the groups they are part of and recolor them.

Some of them are strokes, so they need to be colored differently. But instead of this, you can be smart and use the **Find/Replace** method:

1. To start, identify the objects you want to recolor. What is the common property of these objects? Their color, for example. Click on one of the objects and get the code of that common color – in this case, copy the code of the original turquoise, 2ac1b5, without the # symbol.

2. Now, you need to press *Ctrl + F* to bring up the **Find/Replace** tab on the sidebar. Paste the color code you are looking for into the **Find** field as seen in *Figure 8.4*.

3. Now, get the color code of the new color. In this case, we are looking for orange, with the `ff9955` code. Copy and paste this code into the **Replace** field.

4. As you can also see from *Figure 8.4*, there are quite a few settings and options on this tab. Most of the time, you will only need a few.

5. Under the **Search in** label, select the **Properties** option. This will tell Inkscape that you are not searching for this string of characters in the text but in the properties of the objects.

6. Under **Scope** select **Search in selection**. This is about fine-tuning your search criteria to only find and select the elements you mean to and to avoid randomly selecting and changing parts of the document. For this to work, select all the icons you want to change the colors of. The rest will be out of scope.

7. And that is all! If you click **Find** at the bottom, all of the elements that have a property with the given value will be selected. Now you can recolor, move or even delete them as you need.

8. Or you may select **Replace all**, and the original value will be overwritten with the new one. In this case, all the elements will be selected and given a new color value. For the task at hand, this is the better solution. It does not matter if the color was assigned as a stroke or a fill color since Inkscape now only searches for the value anywhere in the selected part of the XML.

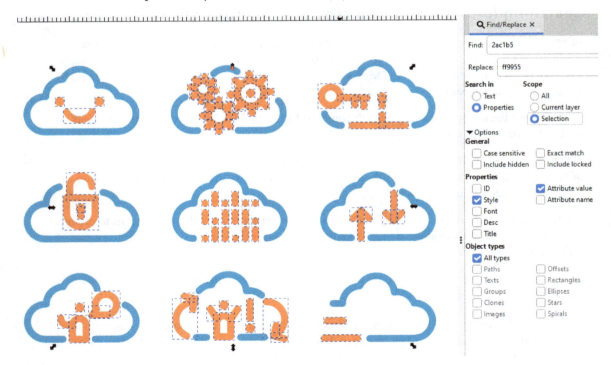

Figure 8.4 – The elements selected using the Find/Replace tab

This simple trick can save a lot of time and work on recoloring multiple elements. The **Find/Replace** tab adds a great deal to the versatility of vector graphics with Inkscape.

Naming and organizing your objects with unique IDs

As mentioned previously, SVG files are built with XML, which means that they are readable for both programs and humans. There is no complex encoding but an easy-to-follow structure of nodes, IDs, and attributes.

You can also use this open structure to your advantage to organize your documents better.

Get any object, path, or group in Inkscape. Right-click the object and select **Object Properties**. Or press *Shift + Ctrl + O* and then select the object.

This will open the **Object Properties** tab, where you can set the ID for the selected object. Every object has to have a unique ID in the document; by default, Inkscape assigns one to them.

Just write the ID to your object or group, that is, `icon001`, `btn002`, `character1_head`, and so on. Finally, click **Set**, and the object is properly identified. Check out *Figure 8.5* for example settings.

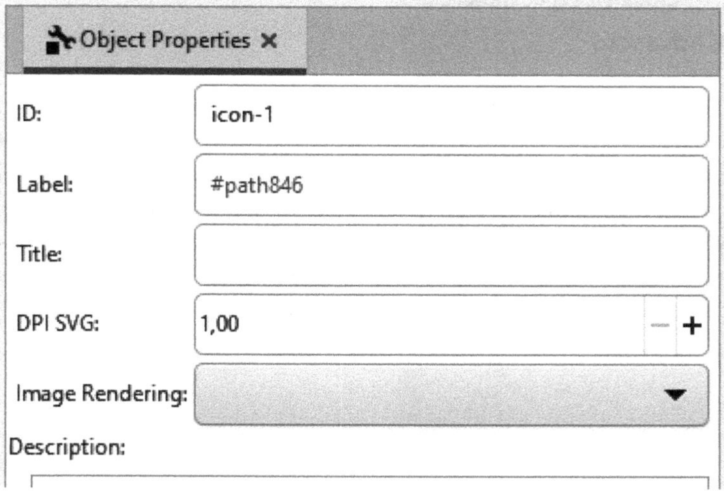

Figure 8.5 – Setting the ID of a single path in the Object Properties tab

So, how can this help you? Do you need to do this for every single object? Of course not, but changing the automatic IDs set by the program makes finding named objects easier. If you want to use the **Find/Replace** tab, you can also check the box next to the **ID** field, so Inkscape will then select all the icons or anything with a specific ID match.

If you are creating game art or working on a complex UI design or illustrations, naming parts of your vector graphic is a useful habit to pick up.

Giving unique IDs to your groups is also helpful when you need to select **Batch export multiple images** as well. This is not a new feature; even the older versions of Inkscape can export multiple selected objects at once (there is a **Batch** export checkbox in the **Export** dialogue in those older versions). There are a few methods for this, and there is a big difference between Inkscape 1.2 and the versions before and one thing in common.

Let's check them out.

- **If you are using an older version of Inkscape**: To start, simply select multiple elements on your page, and check the **Batch export X selected objects** box on the **Export PNG images** tab (to open this tab, press *Shift + Ctrl + E*). This will export multiple PNG files in the defined folder.

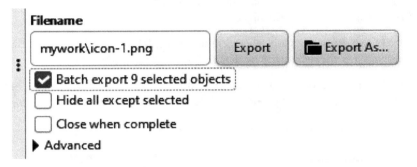

Figure 8.6 – Batch export multiple objects before Inkscape 1.2

- **If you are using Inkscape 1.2 or newer**: Since Inkscape version 1.2 there is a dedicated **Batch Export** tab under the **Export** tab. Here you can export multiple files with different file formats. Not just PNG but SVG, PDF, and JPG are supported. This **Batch Export** tab gives you previews of the objects selected, where you can see exactly what will be exported and with what filename. This solution is much more convenient and helps with your workflow.

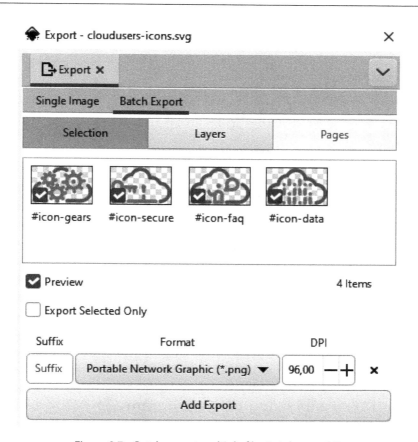

Figure 8.7 – Batch export multiple files in Inkscape 1.2

- **Setting unique IDs**: This is the common point between the two versions. If you did not set your own unique IDs to the objects to be exported, the program will automatically save them with automatic filenames created from the automatically assigned IDs in the SVG. You will have files named path4483.png, g5123.svg, and so on. Then, you should identify them and rename them manually, and for multiple files, that is very time-consuming.

  For example, when you are exporting 50 icons at once, these generated names can be a real drawback since you have to check every single icon yourself and name them according to their content so they can be used later.

But if you did set the IDs in Inkscape for each of the objects or groups to be exported, then Inkscape will save those PNG files with the filenames specified as IDs before. To come back to the icon example, if you give your icons an ID as you go (setting each group a unique ID in the **Object Properties** tab) then Inkscape will assign those IDs as filenames. This will help identify and sort your images and files, later on, sparing you a lot of time in your workflow!

Consider using Inkscape vector more like XML

Even if you are not comfortable thinking about the code behind your graphics, please consider using Inkscape as an accessible editor for your SVG file.

Searching for values in the SVG script is very easy in Inkscape and can save a lot of time and effort. Setting IDs can also be an easy and useful method for you, shortening the time of finding, organizing, and exporting your graphical elements.

In this section, we learned about using IDs for searching and exporting vector elements in Inkscape. Both humans and machines can recognize and read your IDs and other properties in an SVG file. They are used by laser cutters, printing setups, or, as shown in the next section, by Inkscape extensions. Indeed, IDs and specific extensions can be one of the solutions for exporting your images in CMYK!

Overcoming the CMYK color barrier in Inkscape

CMYK has been mentioned before in this book, and users tend to talk about it online as well, but what is the issue here? Inkscape is working with SVG, and SVG is created for the web. It means it is natively running in the RGB color space used by screens and not in the CMYK color space for printing.

Although this issue is less and less important since most printers can do the converting for you if you provide the graphics in RGB, CMYK is still needed if you want to have absolute control over the printed colors. Some printers only accept CMYK formats, and specific files such as PDFs.

Solving the CMYK issue using other programs

In *Chapter 7, Combine Inkscape and Other (Free) Programs in the Design Workflow*, we learned about **Scribus**, which is open source publishing software. In the *Scribus* section, we mentioned one solution for the CMYK challenge: save the vector graphics as an SVG in Inkscape and import the SVG graphics into Scribus.

Then, set the color space and color settings there and export your document into any CMYK color mode and format you want. This way, you can easily generate proper vector graphic files ready to be printed.

You could also use another program mentioned in *Chapter 7, Combine Inkscape and Other (Free) Programs in the Design Workflow*: **Krita**. You can import the SVG as you did in that chapter and export it as a TIFF in CMYK format. A TIFF is a bitmap image, not a vector, but it uses lossless compression. That makes it perfect for printing.

Using the Inkporter extension for CMYK exports

The previous solutions work very well, but for some, it might be too much effort to use a third-party app every single time they need a CMYK vector document for printing.

This is where **Inkporter** comes into the picture. Inkporter is an Inkscape extension that can export parts of your SVG document into several formats! This extension came out before Inkscape 1.2, but it still has some advantages compared to that version.

The new multipage support and the **Extension** tab that was introduced in Inkscape 1.2 are great. But additionally, Inkporter can export multipage PDF documents and JPG and PDF files in CMYK format too!

I mentioned earlier that the IDs would be important later in this chapter. The extension method of Inkporter is ID based. It exports the files and the pages in the PDF booklets based on their IDs with a unique pattern. Let's give it a try:

1. To start, you need to install the Inkporter extension. Close Inkscape if it is currently running. You can download Inkporter from `https://inkscape.org/` under the **Extensions** tab.

2. Installing this extension is simple, just copy the downloaded files into the `Extensions` folder in your Inkscape installation. If you need help, you can follow the installation steps in the description on the download site.

3. After Inkporter is installed, open a document with multiple groups or objects again. The nine icons from the icon set you created during *Chapter 3, Modular Icon Set Design with the Power of Vector* will be the perfect test subject again.

4. As mentioned previously, Inkporter works based on IDs. It will identify the elements you mean to export based on their IDs. Press *Shift + Ctrl + O* to open the **Object Properties** tab and set a unique but repeating ID pattern, that is, `icon-cloud`, `icon-settings`, `icon-load`, and so on.

5. Then head to **Extensions | Export | Inkporter** to open the **Inkporter** popup. Choose a file format; for the icons, SVG will work best. The ID pattern will be recognized if you type in `icon-`, Inkporter will export all the groups that have a unique ID starting with `icon-`.

6. You only need to set an export folder and select **Export**, as shown in *Figure 8.8*, and you are done.

Figure 8.8 – The simple settings of Inkporter

ID patterns are only needed now because we are looking for an effective (lazy) solution by exporting all the icons at once. You could set unique IDs without any repeating pattern for each icon, but then you would have to set them manually for each export.

If you want to export PDF booklets, patterns are even more important since the groups with the patterned IDs will be stitched together into one PDF document.

As you can see, the CMYK issue is still present but is solved by third-party apps such as Krita and Scribus and by the smart Inkporter extension. I really hope that a native CMYK solution will be added to Inkscape in future updates, but until then, these are some working methods to jump over the CMYK color barrier.

The next section is a simple way to make your vector life easier: you will learn how to use and create templates for your Inkscape workflow.

Working faster with custom templates

If you use Inkscape regularly, after a while, repeating the same tasks over and over again is inevitable. You might create one business card after the other or draw icons of the same size every day. This is why, when you create a new document, Inkscape offers to create it from a template to ease your work. There are a number of default templates, including standard paper sizes, the mentioned business card size, and even templates to create seamless pattern designs.

And apart from these, you can create your own custom templates for tasks that you work with regularly. If you know that one task is repeated frequently, and you know you will keep on repeating it in the future, it might be worth creating a template for it. A template will save you time and energy, and it will become an important part of your workflow in no time.

Creating your own custom template

A template file is very simple to design. It is nothing else, just an Inkscape SVG file that is placed in the Inkscape `Template` folder. You might create a general template file; for example, if you create the same size horizontal banners for all websites, create a blank SVG with the page size needed. Or you can create a more specific template for a task you create for a particular client over and over again.

Figure 8.9 – The New From Template popup

To practice, you will create a very specific template: you will design an Instagram post template to use for our practice brand, CloudUsers. Instagram is built around images, but brands and influencers like to post text and quotes in an image format to capture the attention of users.

These posts are all the same size, with a background color, some sample text, and minimal branding. If you have used Canva for Instagram posts, Inkscape might be a more creative and unique solution.

Here is how you can create your own Instagram-ready vector templates:

1. To start, create a new document, and draw a 1080 px X 1080 px square. This is the standard Instagram post size. This will be the basis of our template.

2. Select the square, and press *Shift + Ctrl + R* to resize the page to the selected content. Or use the **Page** tool and resize the page with the resize icon on the **Tool** menu.

3. Now, to the branding and colors. If you created the different versions of the CloudUsers logo in *Chapter 2, Design a Clever Tech Logo with Inkscape*, you should have a white version of it. That version was placed on the gradient we used in the logo. Copy the white logo and the background element with the gradient into your Instagram template.

4. This same linear gradient will be the color of your background; now apply it to the square.

5. As you see, templates can really contain everything a normal SVG file can: shapes, colors, gradients, and even text and guides. To add the guides, draw another square, this time 1000 px X 1000 px. Position it in the center of your page, in the center of the square background.

6. Select this second square and select **Object | Object to Guides** in the top menu. This will turn it into four guidelines, each 100 px away from the edge of your background square. These will help later every time you use this template.

7. Now add some title text using the **bold Montserrat** font to the background. Under that, add smaller sample text, as shown in *Figure 8.10*.

8. Under the text, place the white CloudUsers logo. That's it, you're finished with the design!

Figure 8.10 – A simple Instagram post template design

9. To turn this into a template, simply copy it into the Inkscape `Template` folder at `Share/Inkscape/Templates`. It is in your Inkscape installation folder.

10. If you close and restart Inkscape now, you will see your own Instagram template among the templates offered at the start!

Templates are all about reusing and shaping vector elements. Forget recreating the same things again and again. Build your own template library and make custom templates part of your workflow! In this section, you learned to use templates to start your work faster. In the next section, you will learn about dealing with Inkscape crashes like a pro.

What to do if Inkscape crashes

This part of the chapter could be written about any graphic design program, or about any program really. Nothing is perfect, accidents happen, and programs crash. And sometimes, you just lose the results of hours of hard work.

Inkscape is an open source program, with different versions able to run on various operating systems. People use Inkscape on Windows, macOS, and GNU/Linux, working on computers with very different capabilities.

Inkscape is constantly developed, and new features mean new design possibilities, but they might also cause bugs. And sometimes things are too much to handle, and Inkscape crashes. In the following sections, you will learn what to do on these rare occasions.

Prevention

Your first defense against accidental crashes is to prevent them. Here are a few tips to remember every time you work with Inkscape. Most are useful tips for other design programs as well.

Save your work and save it often

The first thing you should always do right after you draw the first shape is to save your work as an Inkscape SVG file in the project directory. And after that, save your work every time you make a major change. Make this a habit; simply press *Ctrl + S* to quickly save your file. It seems trivial, but I've seen too much work being lost because someone was so deep into their design work that they forgot to save it.

Set up the Autosave timer

If you tend to forget to save, set up **Autosave**! Go to **Edit | Preferences**, and from the **Input/Output** drop-down menu, pick **Autosave**. Here, you can set up how often you wish your work to be saved. You will not notice this feature running, only when you lose your work and find out that Inkscape indeed saved it a few minutes ago. Such a relief!

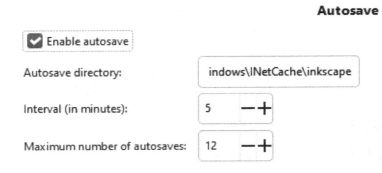

Figure 8.11 – Enabling Autosave in Preferences

Don't overload Inkscape; organize your work

You cannot overload Inkscape, actually, but you can cause it to collapse when you force it to handle too much, and that can overload your computer's memory. When too many objects, too many effects, or too many imported bitmap images are starting to slow down your computer, then it is time to save, stop and clean up your document a bit.

Delete the parts and objects that are not needed anymore, link images instead of embedding them whenever possible, and save new versions of your documents after cleaning them up.

This last tip also helps organize your work, since it is easier to handle multiple smaller files than one huge one for each project. This book itself provides a good example of this method: when designing a whole visual identity, including a logo, an icon set, a website, and illustrations for one company.

It is natural that you copy and reuse elements between the different parts of the project. If you kept these elements in one file, it would be huge, slow, and complicated to find your way around, let alone work with it. This is a good reason to keep your content organized in separate files!

Recover your work

Using the aforementioned preventive tips will reduce the risk of crashing and losing your work. Plus, with every new version, Inkscape is getting more and more stable. But it can still happen that you get an error message and the software simply crashes. And in this case, you need to know how you can recover your graphics in Inkscape.

Emergency backup files

Whenever the program crashes, Inkscape tries to create an emergency file in the same project folder as your original file or in the `User` folder on your computer. It will add a timestamp to the name of the file to easily identify it and distinguish it from the original. Look for the file and check whether it has the correct vector data in it. If it works and has all the graphics you just thought to be lost, rename it to remove the timestamp – just to make things convenient later – and continue your work.

Autosaved backup files

If you have the **Autosave** feature enabled, as shown earlier in the *Prevention* section, then Inkscape creates backup versions of your document automatically. To find these backups, head to **Preferences** again, and in the **Autosave** settings check the path to the autosave directory, as shown in *Figure 8.11*.

Copy the directory location and navigate there, as this is where your backup files are located. Most of the time, they do not have a `.svg` extension, but you can still open them with Inkscape and save them as normal Inkscape vector files.

How to open corrupted files

Sometimes files are saved, but they simply cannot be opened. This might happen without a crash with any program; files just get corrupted.

If you cannot open an Inkscape SVG file, you might try opening it with another program to check whether it really is corrupted. You can try other vector graphical applications such as Illustrator, but if you have none of them installed, modern browsers can even open and display SVG files. Try opening your file in Chrome and see whether it shows the file is empty or damaged.

If the file displays properly, create a manual save with another filename, and try to open it in Inkscape again. You might lose some Inkscape-specific settings this way (such as losing guides), but you might save the bulk of your design and that is what matters.

When you try to open an Inkscape file that worked before, do so with a different version of Inkscape. Most of the time, installing the newest version of the program solves this issue. If not, try a downgrade, or look for an older version – most versions are still available from the official website.

And if a corrupted file cannot be opened with either method, try to look for autosaves and recover a previous version if possible.

Most importantly, remember to save often, enable Autosave, and organize your files because prevention is the best way to keep your work safe!

Summary

This chapter was built differently than the previous ones. In this chapter, you received various tips and tricks that are based on a decade of daily Inkscape usage. First, there was a short reminder about the basic tools and methods and how important it is to know them before digging deeper. Then we learned about LPEs and through a few examples, saw how effective they really are. LPEs provide some of the best automation for vector graphics.

Then we moved even deeper, checking behind the curtain of SVG files and learning about the benefits of using just a bit of XML. The **Find/Replace** function is invaluable when you are working with multiple objects, and object IDs can be used in more than one way too!

CMYK support was always a question for XML, but thanks to the Inkporter extension and other apps, now you know more than one way to solve this problem.

Next, you learned about custom SVG templates and how easy it is to create them in Inkscape. And finally, you prepared yourself and now know what to do when Inkscape crashes and when SVG files get corrupted. Remember: make saving a habit!

The next chapter is the last one of this book, where we will take a look at all you have learned about Inkscape.

9
Conclusion

Which topics did we cover in this book? What type of project did you complete? What did you leave out? Hopefully, this short summary will help you conclude everything you read and learned in this book.

In this chapter, we are going to cover these main topics:

- Conclusions of all the projects
- What's next? Make your own!

Conclusions of all the projects

In the following section, we will go through all the chapters in the book to refresh what you learned. These short conclusions will be reminders about the tasks you completed, the tools you used, and the methods you learned in that chapter. Apart from a simple summary, we will also emphasize the goal of each project.

Chapter 1, Inkscape Is Ready for Work – Design a Business Card as a Warmup!

Most of the first chapter was not focused on a specific project, but on the usability and usefulness of Inkscape. During the chapter, we showcased a handful of new tools and features to show how Inkscape matured into being a professional tool.

Because the first chapter was mostly theoretical, it contained only a short warmup project: designing a simple modern business card. The role of this exercise – apart from being a warmup for the style of this book – was also to assess your level of Inkscape.

Chapter 2, Design a Clever Tech Logo with Inkscape

This chapter started with a small amount of logo design theory, with thoughts about simplicity, creativity, and message. Then you started to create a real logo for a made-up tech company called CloudUsers.

It started with planning and sketching, and how to import your sketch into Inkscape. Then you moved on to designing in Inkscape, focusing first on shape, and then on color. Shapes were created with the **Node** tool (**Path editor** tool) and Boolean operations, while the colors were implemented via gradients.

You also learned about how duplicating helps the creative process, before starting to add text to your logo design. Finally, you created different logo versions and exported them into different formats. At the end of the chapter, we also learned about the most common formats used for logo design.

The goal of this chapter was to give you a step-by-step guide you can follow to create your own logos in Inkscape!

Chapter 3, Modular Icon Set Design with the Power of Vector

When the logo was finished, the fictional company, CloudUsers, remained with you for this project as well. This time, you created a whole set of icons in the style of the previously designed logo. After a bit of theory again, you created the first icon of the set, and with it laid down the rules to follow for the rest of the icons: colors, strokes, and style.

The longest – and most creative – part of the project was to design the remaining icons from this modular set of rules. While creating them one by one, you practiced working with the **Bezier** tool, modifying paths with the Node editor, and organizing the icons with groups.

Then, after all the nine icons were finished, you learned about exporting multiple icons to different formats.

This project was geared towards UI designers and aimed to set you up for modular thinking while designing with Inkscape. Set your rules, create your building blocks, and repurpose and reuse elements creatively.

Chapter 4, Create Detailed Illustrations with Inkscape

Creating the illustration was one of the longest projects covered in the book. It needed to be long because we went into more detail than with the icons and the logo design projects since we were trying more tools and methods.

The project in this chapter focused on the workflow of creating a business illustration with many details. The complexity of an illustration is only dictated by the details added by the illustrator. And where there are a lot of details, there are a lot of chances to develop your skills.

You started with sketching in Inkscape, an easy way to focus on ideas and form before everything else. Then you built the basic color palette for the illustration. After this, you drew your first character, based on the character shapes in the logo.

You learned the *Sandwich method*, an easy way to create shapes that perfectly match the edges of the shape under them. Then you added two more characters and added a cloud and a giant laptop as background elements.

You colored the elements and the characters, both applying the original color palette and mixing new ones using the **Color picker** tool. Finally, you added shading to the characters and background objects, adding another level to the illustration.

Chapter 5, Edit a Photo and Create a Hero Image in Inkscape

The aim of this project was to shed light on the *photo editing* capabilities of Inkscape. During the project, you created a hero image, an image banner that can be used on websites or social media.

In this typical graphic designer task, you first practiced creating a depth-of-field effect in Inkscape using Blur, Clip, and Mask. You learned that within limits, Inkscape is good for covering and retouching parts of bitmap images and photos. Then you repurposed graphics from the icons you created earlier to add a network illustration over the photo, and after that, you learned a simple trick to draw *into* the photo using Clip and saw how to trace parts of the image.

Chapter 6, Flexible Website Layout Design for Desktop and Mobile with Inkscape

The project in the sixth chapter was based on what you created earlier, and so it helped you hone real-life designer skills. You had to create a website design using the logo, the icon set, and the illustration you drew in the previous project.

To achieve this, you first created a simple wireframe layout of the website, relying mostly on the **Rectangle** tool, alignment, and guides. Here, you set all the building blocks of a simple contemporary website: the position and size of the menu, the header, the product cards and icons, and the footer.

Then you *only* had to assemble the desktop version of the site, using the layout with the colors, fonts, and vector elements you created earlier. After this, you had to realign and reassemble the desktop website design into a mobile version of itself.

The goal of this project was not only to show you how to create a basic website design, but to focus on the flexibility of vector design too. Designing something new and coherent with the given brand elements is a common task every designer has to tackle.

Chapter 7, Combine Inkscape and Other (Free) Programs in the Design Workflow

This chapter took the focus away from working solely in Inkscape and instead looked at combining it with other programs. The goal? To use Inkscape as a part of your real-life workflows and make them more effective. Using the power of the SVG format, you followed three different applications for three short projects:

- You first learned how to enhance your vector illustration in Krita. Adding textures and soft shadows with freehand raster brushes is something that can really elevate a simple vector image.

- Then you used Scribus to create a flyer using your SVG illustration and the logo you created earlier. Scribus is also a great tool to export your Inkscape creations in CMYK format; you practiced using that tool as well.

- Finally, you took a short tour of Blender, a 3D modeling and animating tool. You used a simplified SVG version of the CloudUsers logo and learned how to turn it into a digital 3D object.

Chapter 8, Pro Tips and Tricks for Inkscapers

This chapter was not a single project to follow step by step. This chapter was about situations you might encounter using Inkscape, and about the tips, solutions, and workarounds for those situations. First, we took another look at the basic tools of Inkscape and saw what you need to learn and practice to get strong vector skills.

Then you learned about the usefulness of **Live Path Effects** (**LPEs**), a set of tools that help you save time and effort while modifying your paths in real time. LPEs are really helpful, so it is wise to try more than the two we examine in detail during the chapter!

After that, we looked behind the scenes and utilized the SVG format Inkscape works in. This XML-based format is easy to navigate. You learned how to find and replace the properties of multiple vector objects at once.

Then we looked at another solution for the CMYK export problem, this time using Inkporter, an Inkscape extension. The next tip was wrapped in a small project about creating and using custom Inkscape templates. You practiced this by designing an Instagram post template.

In the final tip, we addressed the dreaded moment of Inkscape crashing. You first learned about preventing losing your work, then about recovering it, if an accident occurs.

This chapter is a collection of tips that help you to use Inkscape every day as your main design tool.

What's next? Make your own!

After completing all the projects in *Inkscape by Example*, start implementing these projects in your own design work. But the question remains: *what's next?*

Copy, try, learn, and then make your own versions of the projects. That provides a better understanding than just following the projects or simply reading through the chapters.

The final conclusion is this: after creating the advanced projects, you will be able to tackle any task related to Inkscape. You will have a design blueprint you can look into, whether it's logo design, icon design, or illustration. Each chapter was written in a way to make them easy to scan through if you are looking for a special solution that was explained during that project.

This could be the slogan of this book: *make your own!* As you can see, there is more to Inkscape than just learning the tools. You need to practice and build your own projects, modify the examples of this book, and create your own designs!

Open Inkscape, continue practicing, make your own!

Index

Packt.com

Subscribe to our online digital library for full access to over 7,000 books and videos, as well as industry leading tools to help you plan your personal development and advance your career. For more information, please visit our website.

Why subscribe?

- Spend less time learning and more time coding with practical eBooks and Videos from over 4,000 industry professionals

- Improve your learning with Skill Plans built especially for you

- Get a free eBook or video every month

- Fully searchable for easy access to vital information

- Copy and paste, print, and bookmark content

Did you know that Packt offers eBook versions of every book published, with PDF and ePub files available? You can upgrade to the eBook version at packt.com and as a print book customer, you are entitled to a discount on the eBook copy. Get in touch with us at customercare@packtpub.com for more details.

At www.packt.com, you can also read a collection of free technical articles, sign up for a range of free newsletters, and receive exclusive discounts and offers on Packt books and eBooks.

Packt is searching for authors like you

If you're interested in becoming an author for Packt, please visit `authors.packtpub.com` and apply today. We have worked with thousands of developers and tech professionals, just like you, to help them share their insight with the global tech community. You can make a general application, apply for a specific hot topic that we are recruiting an author for, or submit your own idea.

Share your thoughts

Now you've finished *Inkscape by Example*, we'd love to hear your thoughts! Scan the QR code below to go straight to the Amazon review page for this book and share your feedback or leave a review on the site that you purchased it from.

`https://www.amazon.com/dp/1803243147`

Your review is important to us and the tech community and will help us make sure we're delivering excellent quality content.

Download a free PDF copy of this book

Thanks for purchasing this book!

Do you like to read on the go but are unable to carry your print books everywhere?

Is your eBook purchase not compatible with the device of your choice?

Don't worry, now with every Packt book you get a DRM-free PDF version of that book at no cost.

Read anywhere, any place, on any device. Search, copy, and paste code from your favorite technical books directly into your application.

The perks don't stop there, you can get exclusive access to discounts, newsletters, and great free content in your inbox daily!

Follow these simple steps to get the benefits:

1. Scan the QR code or visit the link below:

https://packt.link/free-ebook/9781803243146

2. Submit your proof of purchase

That's it! We'll send your free PDF and other benefits to your email directly.